The Geographer's Art

As professionals we claim only that we are privileged to devote ourselves to the field of geography. Neither we nor our academic predecessors discovered the field, nor have we been the only ones who have tilled it ... We have never been more than a few of those who contributed to the growth of geographic knowledge. The interest is immemorial and universal; should we disappear, the field will remain.

Carl Ortwin Sauer, *The Education of a Geographer* (1956)

The
GEOGRAPHER'S
ART

PETER HAGGETT

Basil Blackwell

Copyright © Peter Haggett 1990

First published 1990

Basil Blackwell Ltd
108 Cowley Road, Oxford OX4 1JF, UK

Basil Blackwell Inc.
3 Cambridge Center
Cambridge, Massachusetts 02142, USA

British Library Cataloguing in Publication Data

A CIP catalogue record for this book is available from the British Library.

Library of Congress Cataloging in Publication Data

Haggett, Peter
  The geographer's art/Peter Haggett.
    p.   cm.
  Includes bibliographical references.
  ISBN 0–631–17144–4
  1. Geography – Philosophy. I. Title.
G70.H22 1990   89–28937
910'.01–dc20   CIP

Typeset in 10 on 12 pt Sabon by
Wearside Tradespools
Printed in Great Britain by
T. J. Press Ltd., Padstow

*For Jackie – the third of my four
non-geographical children – with love*

Contents

Figures

CHAPTER 4

CHAPTER 5

CHAPTER 6

Plates

Plates are located between pages 110 and 111 in the text.

Acknowledgements

The author and publishers wish to thank the following who have kindly given permission for the use of copyright material:

Association of American Geographers for figs 1.4, 6.4; Academic Press, Inc. (London) Ltd for figs 3.4, 4.5; Edward Arnold for figs 2.8, 4.4; The Australian National University Research School of Pacific Studies for fig. 5.1; Basil Blackwell Ltd. for figs 2.7, 2.10, 3.3, 5.5; Cambridge University Press for figs 3.6, 5.6; Gerald Duckworth & Co. Ltd for fig. 1.7; Economic and Social Research Council for fig. 4.9; The Geographical Association for fig. 7.3; Gustav Fischer Verlag for fig. 4.1; Torsten Hägerstrand for fig. 6.2; Hakluyt Society for fig. 2.4; Institute of British Geographers for figs 3.9, 6.3; New Zealand Geographical Society for fig. 4.10; Monash University Department of Geography and Environmental Science for fig. 7.9; Pion Ltd for figs 3.2, 3.7, 3.10, 7.11; Annette Reenburg for fig. 1.5; Royal Geographical Society for fig. 3.1; Society of Economic Paleontologists & Mineralogists for fig. 2.9; Rowland R. Tinline for fig. 5.4; University of Bristol Department of Geography for figs 5.10, 6.6; University of California Berkeley Department of Geography for figs 4.2 and 8.6; University of Cambridge Department of Geography for fig. 4.2B; University of Illinois Press for fig. 1.6; University of Minnesota Press for fig. 8.4; Whitcoulls Ltd for fig. 1.1; John Wiley & Sons Ltd. for fig. 2.10.

Preface

It's perhaps right for a book about geography to have been started by a frozen lake in North America and finished on the edge of the tropical rain forest in Borneo. What is wrong is the six-year gap between the two events, when little, if anything, was added. Let me try to explain how this small book came to be written and why it took so long.

I

I began to write the first of these eight essays in the late spring of 1983. The Wisconsin snow flurries were scudding across the square from the old University Club on Murray Street so that the outlines of ice-bound Lake Mendota were blurred and indistinct. Both the time and the place were symbolic. My fiftieth birthday was past and I was in a nostalgic mood, missing my family and finding excuse after excuse for not putting pen to paper.

Wisconsin brought back memories of my undergraduate years at Cambridge. It was Richard Hartshorne's book that I had first bought as a freshman in October 1951. I recall buying it at the old, crowded Heffers bookstore in Petty Cury and cycling back along Grange Road to my lodgings; on that crisp autumn afternoon with turning leaves and bonfire smoke, the austere grey-covered volume of *The Nature of Geography* bounced in the wicker basket on my cycle.

My first efforts to understand the book that evening did not bring much success. As the term wore on I slogged my way over the opening foothills and tried some of the higher paths in the later chapters. No relevant examination questions were set at the end of the year and, newly enthused by geomorphology and down on funds, I confess to selling it to buy von Engeln's *Geomorphology*. But the Hartshorne book had an enduring effect on my thinking and before another year was out I had to buy my second copy.

On graduation it was touch and go as to whether I went to Wisconsin to follow up these leads. The decision to stay on at Cambridge had something to do with academic matters, but was much more influenced by the Homerton College girl with the sparkling eyes. In the event it was my Cambridge room-mate Ken Warren who came to Madison on a Whitbeck Scholarship, while Gerald Manners headed for Indiana and Michael Chisholm for Oxford; like Peter Hall, I stayed on in Cambridge to do research. The five of us, all Gus Caesar's students, were infected for life with the geographical virus.

All that had happened thirty years before and so much time had slipped by that — even on optimistic actuarial reckoning — less productive time must lie ahead. The opportunity of a brief leave period funded by the Leverhulme Trust had allowed me to come to Wisconsin. There I felt it was high time I tried to put down my thoughts about the subject that has dominated my working life before the opportunity or the willpower slipped away.

In the end the grand book became this small set of essays. I would like to have painted on a larger canvas but like Mussorgsky (and John K. Wright) I found I could only present some 'Pictures at an Exhibition'. All the pieces begun at Madison were newly painted but, as some were painted over old canvases, some familiar themes may show through here and there.

II

That part of the preface was written six years ago and I've had to blow the dust off it in a different climate. The spring of 1989 saw the second of several visits to the little town of Bandar Seri Begawan on the western coast of Borneo. Again the real reason for my being there was a Geography Department; this time at the little university which is now growing strongly. Long air journeys and delays suggested I should bring the chapters with me and try to complete them.

The six-year delay between the Madison start and the Bandar finish was due to a telegram which reached me in the autumn of 1983 when I was working in Fiji. It asked me to take over the Vice Chancellorship (President would be the American term) of my home university for a bridging period between an outgoing and an incoming knight, one a nuclear physicist, one a mathematician. These were difficult days for universities in Britain, with heavy cuts in funding from the government body to cope with and no time for writing or reflection. It was a busy and interesting year.

My expected return to geography was further put off when, gamekeeper turned poacher, I was appointed by the Secretary of State for Education to join Britain's sometimes praised, sometimes reviled, funding body, the

University Grants Committee, and to chair several of its critical panels. Whitehall further bound me by an appointment to Britain's radiation watchdog, the National Radiological Protection Board, which after the Chernobyl incident greatly increased its intensity of business. So I spent the years between Madison and Bandar largely out of academic geography and much concerned with national questions of university finance.

III

So this year offers a brief breathing space in which to complete these essays before a new research project for the Wellcome Trust is started. My only other choice is to put them permanently into the archives along with the other unpublished and unpublishable papers.

As I reread them, trying to make the decision, I'm struck that they are personal and partial; that they give only one man's view of a subject which is rich and complex. Too rich and complex for any short book to encompass.

So I've simply tried to say why I find this subject of geography so fascinating. This is the point – fascinating rather than important or relevant (though it is these too) – which I would want to get over to the reader. For, like Carl Sauer, I think an interest in geography is immemorial and universal. Most of us surely will be touched by it at some time just as we are touched by music.

Music of some form – whether orchestral, choral, folk, religious, supermarket muzack – is with us much of the time; it soothes us, stimulates, angers. Languages wrap around us in a similar way. Landscapes, environments and regions of the earth's surface also bring a surprisingly sharp set of responses. We are moved to tears by a great mountain range, awed by the vastness of a wilderness area, angered by the devastation of a blighted city area, or intrigued by global waves rippling into local regions.

But what separates out the scholar – whether studying music, language or geography – from the rest of the population is the approach to these universals. Each is privileged to study the structure, the grammar and syntax, of the forms which they observe. Geographers are concerned to find the pattern and structure and meaning that lies in the world's regional diversity at all scales, just as musicologists are concerned with musical order and composition.

Now the study of geography at university will give scientific insights, will provide skills in increasing demand in the world of work, will provide a philosophical framework for environmental and global problems. But I think its most precious gift is none of these. Geographical study can enrich all the remaining years of one's life so that what is seen is understood and appreciated. This is the heritage that we are privileged to pass on to the next generation.

IV

Finally I must express my thanks to those who encouraged me to write this book, and helped me to complete it. I've been fortunate in my academic 'climbing companions' over the years, and I owe a particular debt to Richard Chorley and Andrew Cliff with whom I've worked so closely for so long. We have worked together on a dozen volumes and scores of research papers and many of the sensible ideas here are theirs rather than mine. The more purple passages they would gladly disown.

At Madison where the book's keel was laid down my list of debts is as long as the Faculty List in Science Hall. Bob Sack translated an interest into the reality of a visit and his welcome, and that of his colleagues, has been in the true tradition of mid-western hospitality. I owe a special debt to Richard Hartshorne for long hours of discussion and debate. If we did not always agree then we have always worked patiently to see what separated us. I share the debt all geographers owe him.

At Bandar where I finished the book I am grateful to Goh Kim Chuan and his colleagues in the Geography Department at the University of Brunei Darrusalam. They taught me, as have their colleagues at Singapore and Kuala Lumpur, that there are other valid perspectives and world views of geography than just those current in Europe and North America.

My time away from Bristol would not have been possible but for the generosity of the Leverhulme Trust and I'm grateful to Ronald Tress, then its director, for guiding a late application and for not pressing for any quick report. Michael Morgan took on the headship of the Geography Department at a difficult time in the life of the University and threw away my forwarding address for the vital Wisconsin months.

Since then John Thornes has been the most encouraging and supportive head who has, typically, led from the front in reinforcing a strong research tradition in the department. My Bristol colleagues Nigel Thrift and Allan Frey took time off to read through the manuscript and drew my attention to many errors and missed opportunities: so too did two anonymous readers, one from each side of the Atlantic. I'm grateful to all four readers and, while I've tried to correct the errors and grasp the opportunities, no doubt plenty of both will have eluded me. Valerie Shepard had the unenviable task of keeping my English within reasonable bounds and if the book proves readable much credit will be hers. John Davey at Basil Blackwell, the most imaginative of publishers, took a personal interest in the manuscript and grasped it from me when I would willingly have put it back in the drawer for another six years. Critics will have their own views on who should have won that tug of war.

The greatest debt for finally completing this work must be to Margaret Reynolds, who typed and retyped the final version. As my secretary and mind-reader for twenty-three years, she has lived with a diary in which things hoped for have always triumphed over the evidence of things gained. With her retirement this month chaos will rapidly replace any semblance of order to which my office ever pretended. We both share the view that I have now written far too much and should join a contemplative geographic order.

As for the 'Homerton College girl' I described earlier, she remains the compass by which I steer. Oskar Spate, the dean of Australian geographers and a member of my old Cambridge college, put the public and private life of a geographer better than my words can. His aim, he said, was simply 'to make what one can of this our earth, and if the cosmic order holds nothing for us either of despair or hope, to find our happiness in social duty and private love.' I have been blessed in both.

Chew Magna   PETER HAGGETT
Autumn 1989

1

A Distant Mirror

Of course the first thing to do was to make a grand survey of the country she was going to travel through. 'It's something very like learning geography,' thought Alice, as she stood on tiptoe in hopes of being able to see a little further.
Lewis Carroll, *Through the Looking Glass and What Alice Found There* (1872)

Goodness knows what possessed me to become a geographer. Was it some curious alchemy of DNA or some critical experience of childhood?[1] Reading Conrad or *The Riddle of the Sands* at too early an age? Or being given battered 1920s *National Geographics* when recovering from a scarlet fever attack? Or was it simply a built-in curiosity to wonder what lay over the distant ridges of the Quantock Hills (west) and Mendip Hills (east) that bounded my attic bedroom horizons?

Certainly I can remember no Pauline conversion on a road to Damascus nor a Wesleyan warming on a London street. But I can recall precisely when the final die was cast and the odd circumstances that surrounded it. At the age of sixteen I was stupid enough to injure myself in a school rugby match. This set off an old problem and, as a result, I had to spend several months in an orthopaedic hospital before and after a spinal operation. For the initial period I lay encased, pharaoh-like, in a long head-to-toe plaster cast, so that any movement was restricted. The only way I could get to read a book was through a set of mirrors. By reversing a book and laying it flat on my chest the pages could be reflected, the right way up, to a final mirror just above my head. Other adjustments allowed a view of the lawn and trees outside the ward windows.

For that short period of my life the mirrors were critical and, ironically, they were to play an indirect part in determining my future course. For public school examinations (the old Higher School Certificate) were only a year away and although in hospital I was not let off normal school work. But reading through the mirrors was tiring and the first 'serious' book I was brought and whose illustrations captured my attention was Charles Cotton's *Geomorphology of New Zealand*.[2] At the time I was uncertain about

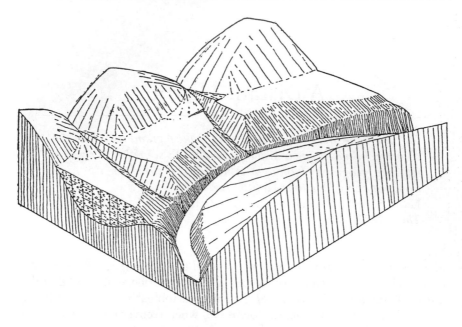

Figure 1.1 Landscape in a looking glass. Cotton's sketch of the terraces of the Shotover River, South Island, New Zealand, showing a river valley with multiple cycles of erosion and deposition. Compare with plate 1.
Source: C. A. Cotton, *Geomorphology of New Zealand*, Christchurch: Whitcomb and Tombs, 1942, fig. 235, p. 232.

which subject I would study at university or, indeed, whether I would be fit enough to go on to university at all. Cotton's diagrams of finely-etched landforms (see the example in plate 1 and figure 1.1) and the problems which their morphology and origin posed made up my mind. I resolved to take forward the study of 'physiography' by whatever route was open to me.

By the age of eighteen I had been lucky enough to get an Open Exhibition to a Cambridge college, St Catharine's, and the memory of that period in an orthopaedic ward was fast fading. But the memory of the mirrors kept cropping up. The book's author, by then Sir Charles Cotton, came to the Geography Department in Downing Place and lectured on New Zealand landforms in the geomorphology programme. Thirty years later I was to go to New Zealand and see the great gorges of the Buller and Waimakiriri at first hand, the reality as sharp then as when first glimpsed in a hospital ward mirror.

THE MIRROR AS METAPHOR

Mirrors and their looking-glass worlds are a well-polished metaphor. Barbara Tuchman wrote a marvellous book about the fourteenth century called *A Distant Mirror*.[3] Tuchman's title underlines the fact that she is using the history of that distant century as a mirror to reflect light and understanding on parallel experiences of the twentieth century. The earlier period, although one of great confusion and turbulence, still managed to produce much of enduring value in poetry and buildings: it was the century of Chaucer and William of Wykeham. Barbara Tuchman's mirror shows us the links between the two centuries, each period marked by massive human suffering and confusion, but each with brave and gifted individuals who still reach outside their own time, even into following centuries, with encouragement and inspiration.

But mirrors can also be used in a forward-looking sense. The physicist Freeman Dyson takes up Tuchman's metaphor and uses it to look forward rather than back.

> The future is my distant mirror. Like her [Tuchman], I use my mirror to place in a larger perspective the problems and difficulties of the present. Like her, I see in my mirror great panoramas of human suffering and turmoil. But that is not all. I also see individual human beings to whom future generations will be grateful for the heritage which they have left to us.[4]

Dyson goes on to argue for a scientific vision of the universe as a harmonious whole in which time past and time future have no absolute existence. For him, as for Einstein, the distinction between past, present and future is only a 'stubbornly persistent illusion'.

If the historian uses mirrors to look back and the physicist uses mirrors to look forward, then the geographer's use of the mirror analogy lies in a different dimension – that of space. Thus the Minnesota historical geographer, Ralph Brown, used *Mirror for Americans* as the title of one of his studies of European settlement of the Atlantic seaboard.[5] Here the mirror was used in a sideways manner, reflecting the experience of European landscapes into the understanding of settlement along the Atlantic seaboard.

I wish to use Ralph Brown's sense of mirror in these essays but to intertwine it with two others. Basically I am trying to explore geographical space as Barbara Tuchman explores historical time, with place, rather than period, reflected in my mirror. An example of this kind of approach is given

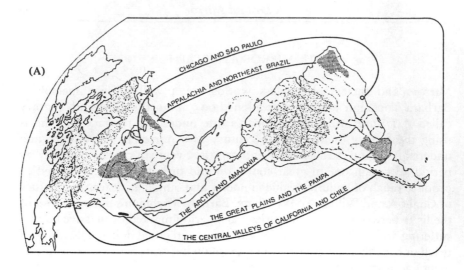

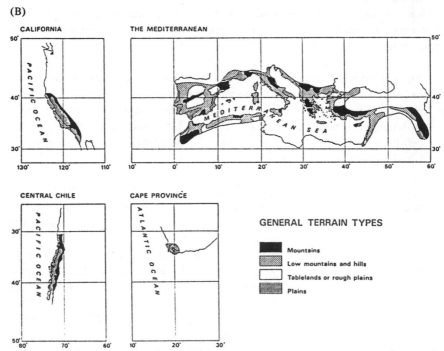

Figure 1.2 Regional comparisons as a mirror. (A) Analogues of five geographical regions in the United States and Canada used to throw light on five comparable regions in Latin America. (B) Terrain types for two of the American regions, California and Central Chile, in relation to other areas of Mediterranean climate worldwide. Note that Australian areas omitted.

Source: A. H. Siemens, *The Americas: A Comparative Introduction to Geography*. North Scituate, Mass.: Duxbury Press, 1977, fig. 1.11, p. 15; Norman J. W. Thrower and David E. Bradbury, *Chile–California Mediterranean Scrub Atlas: A Comparative Analysis*. Stroudsberg, Penn.: Dowden, Hutchinson and Ross, 1977, plate III, p. 7.

in figure 1.2. It is taken from an introductory text on the regional geography of the Americas which tries to familiarize students from North America with the diversity of environments and cultures in Latin America.[6] It does this by matching five 'known' regions in the United States and Canada with five 'unknown' regions south of the Rio Grande. Don Quixote admonishes us, 'all comparisons are hateful' and the comparative method has its drawbacks. But done carefully, with attention to nuances and to similarities as well as differences, regional simile and metaphor can prove useful.

Secondly, I stress distance in the chapter title as a proxy for certain aspects of space itself. For distance has two geographic meanings: distance in the lateral sense of the space separating the objects in the mirror in the plane in which they exist, and distance in the vertical sense of the observer's distance from that plane. From lateral space come notions of cartography, morphology and locational analysis; from vertical space come ideas of scale. If it is held close, fine detail can be seen; if far away, then only the broad outlines.

Thirdly, I stress mirror in the title as indicating the remote view which geographers take of the world, a remoteness forced on them by the size of the objects they study. The image seen in their mirror is not the global reality itself but a symbolic reflection of that reality in which, grind and polish the surface as we may, the image remains an image only.

The emphasis on space and on geometry in these essays is a personal one, though I hope not wholly a quirky one. Throughout my academic life – and certainly in writing this book – I have found the spatial nature of geography its most fascinating side. Yet a reader must resist the temptation to mistake the necessity for a spatial dimension in geography for its sufficiency. My mirror catches only part of the geographical image, and there are other important themes.

MAPS AS MIRRORS

But what exactly do we mean by a 'geographical mirror'? I shall argue that its essential structure is spatial and that its most common (though not exclusive) form is the map.

The emphasis on mapping throughout these essays is largely for historical accuracy rather than for philosophical argument. As Richard Hartshorne's *Nature of Geography* showed so clearly, what geographers actually do may differ from what they say they are doing[7] and the actual form of the discipline may be very different from what its practitioners preach. I have spent more time than I care to admit turning thousands of pages of writing by geographers trying to see the consistent themes that emerge and that give it its special character. And although we say we are doing many things, in most cases geographers at work are often simply generating and dissecting – encoding and decoding as the cartographers say – map after map. (As we shall see later 'simply' may be a misleading description.)

This conflict between the problems we wish to solve and those which actually lie within our capability is illuminated by the exchanges between Arthur Koestler, the critic, and Peter Medawar, the biologist. Koestler finds it difficult to understand why scientists so often seem to shirk the really fundamental or challenging problems. For example, he wonders why the 'genetics of behaviour' should still be uncharted territory and wonders if the framework of Neo-Darwinism is too rickety to support such an enquiry. Medawar patiently responds.

> The real reason is so much simpler: the problem is very, very difficult. Goodness knows how it is to be got at. It may be out-flanked or it may yield to attrition, but probably not to direct assault. No scientist is admired for failing in the attempt to solve problems that lie beyond his competence. The most he can hope for is the friendly contempt earned by the Utopian politician. If 'politics' is the art of the possible, research is surely the art of the soluble. Both are immensely practical-minded affairs.[8]

So, in this book, I argue for a practical and pragmatic approach to geographical practice. For if science is the art of the soluble, then much of geography is the art of the mappable.

But what do we mean by mappable? A map may be simply defined as a representation of the earth's surface, or a part of it, showing its physical or human features, delineated on a flat surface of paper (or other materials) according to some set of rules. The Oxford English Dictionary records its first use in this sense in 1527. Certainly by 1610 Shakespeare was using the term: 'I am near the' place where they should meet, if Pisanio have mapped it truly'.[9] The notion of 'truth' in maps to which Shakespeare refers relates to the set of rules by which the map is constructed. Over the years, both the noun and the verb have changed their meaning. 'Map' is widely used in its figurative sense to denote the relative importance of a person or place; to be 'on the map' or 'off the map' was an indication of standing. At the other extreme, mapping has acquired a precise mathematical significance.

MAPS AS TOUCHSTONES

So central is the map to geographical practice that some observers suggest it can form a diagnostic or touchstone for determining whether a work is 'truly geographical'. For Richard Hartshorne in *The Nature of Geography*, the role of the map within geographical writing is universal and its centrality unchallenged:

> So important is the use of maps in geographic work that ... it seems fair to suggest to the geographer a ready rule of thumb to test the geographic quality of any study he is mapping: if the problem cannot be studied fundamentally by maps − usually by a comparison of several maps − then it is questionable whether or not it is within the field of Geography.[10]

The example used by Hartshorne to illustrate his assertion is the contribution made by the American geographer, Isaiah Bowman, to the study of boundary problems and tension zones at the Versailles Peace Conference after World War I.[11] Maps were also considered by George Washington to be essential in rebuilding after the War of Independence, and he called for 'Gentlemen of known Character and probity' to be employed in making them.[12]

Although Hartshorne and another leading American geographer, Carl Sauer (plate 2), have been thought to take different views of many issues in geographical philosophy, they were at one on the central role of maps. On pondering why people are attracted towards geography, Sauer writes:

> May a selective bent towards geography be recognized? The first, let me say, most primitive and persistent trait, is liking maps and thinking by means of them. We are empty-handed without them in the lecture room, in the study, in the field. Show me a geographer who does not need them constantly and want them about him, and I shall have my doubts as to whether he has made the right choice of life. The map speaks across the barriers of language.[13]

The use of the term language is interesting for, as Sauer goes on to expound, maps may themselves be regarded as a special kind of international language (or set of international languages) for describing the earth and its regions.

If maps are such a powerful indication of geographic interest and aptitude, how do we square this with the use of maps by scholars in other fields? Maps may be critically important to the geologist or the civil engineer, and of more than passing interest to biologists or to historians. Clearly we can make maps to show the distribution of phenomena from aardvarks to zygotes, and even the most broad-minded or empire-building of geographers would not over-extend his disciplinary boundaries to include quite so wide a range of phenomena. How far then is the map a distinctively geographic mirror?

If we regard the boundaries of academic disciplines as convenient markers of the limits of focused concern and competence (rather than God-given divisions of reality), then such questions are secondary. But these still merit a reply and a response must be two-fold. First, maps play a distinctly more prominent and central role in geography than in other disciplines. No other insists that students include courses on map making, map reading, map projections and the like in their core curriculum. Second, other disciplines turn to geographers for the production of maps, for their collection and care, and for their interpretation and analysis. It is relevant to recall that 'Geographic Information Systems' (GIS) was coined by computer scientists to describe the software they had invented to link and map data for small areas. Thus both internal views (by geographers) and external views (by non-geographers) reinforce a considerable but not exclusive concordance between geography and mapping.

DEFINING GEOGRAPHY

Even if mapping and spatial structures are necessary components of geographical understanding, they are certainly not sufficient. Geography has more widely-spread roots in its past and its present objectives are also much broader.

We can see this most readily by looking at how the field is defined. For me, geography is most succinctly described as 'the study of the Earth's surface as the space within which the human population lives'.[14] As the word comes from the Greek *geo*, the earth, and *graphein*, to describe and write, geography could be loosely construed as simply writing about the earth. But historically its use has become more specialized.

The best known and most widely used formal definition of the field was provided by the American geographer, Richard Hartshorne, in his *Perspectives on the Nature of Geography*: Geography is concerned to provide accurate, orderly, and rational description and interpretation of the variable characters of the Earth's surface.[15] One difficulty with this definition for the layman is that the last two terms need some elaboration.

By 'variable character' geographers mean the spatial variation that may occur at all map scales, from the globe itself (say the variations in elevation between one continent and another) down to very local areas (say the windward and leeward sides of a small lake). Although the word 'character' was originally confined to the visible features of the landscape, this is now more broadly drawn. Sauer's essay on *The Morphology of Landscape*[16] is now more than half a century old but the parallels it drew between geographers' study of the natural landscape (*Urlandschaft*) and the cultural or man-made landscape (*Kulturlandschaft*) remain helpful.

By 'Earth's surface' is meant the thin shell, only about one-thousandth of the planet's circumference thick, that forms the habitat or environment within which the human population is able to survive. So in practice, we are talking about a stage which has an area of half a billion square kilometres (70 per cent of it water covered) and with a thickness of up to around 10 kilometres.

But the critic may object here that, as defined above, geography occupies a very puzzling position within the traditional organization of knowledge. I agree. It is neither a purely natural science nor a purely social science. Its intellectual origins as a distinctive field of study predate such a separation, going back to classical Greece, when man was viewed as an integral part of nature. In that period the geography of any area would be written so as to include descriptions of both the animate and inanimate things found there. As we shall see in chapter 6, although individual scholars wrote geographical descriptions over the ensuing centuries, and although geographical societies flourished from the early nineteenth century onward, geography established itself as a university discipline rather late. (Separate departments of geography had emerged in German-speaking countries in the 1870s, but not generally until the present century in the case of Great Britain and the United States.) By then the division of academic studies into the natural sciences on the one hand and the humanities and social science on the other had already become crystallized in formal faculty organization.

In practice, therefore, geography had to be fitted, albeit somewhat awkwardly, into an already established order of organized knowledge. Sometimes it found itself part of a natural science faculty, sometimes in humanities or social science faculty, and sometimes divided between the two. Such ambiguity extends right up to national level with Geography being recorded as both a physical science and a social science in British *University Statistics*[17] but arbitrarily allocated to the social sciences by the University Grants Committee for funding purposes. In consequence rather powerful external forces (as well as an internal logic) have tended to split geography into two parts: a geography of the natural world termed 'physical geography', and a geography of the manmade world termed 'human geography'. This pressure has sometimes been sufficiently strong, as in some Swedish universities, to lead to the establishment of separate departments of physical geography and human geography with rather tenuous cross-links.

While the short-term advantages of this separation accrued in terms of the integration of one part of geography with its neighbouring science (e.g. geomorphology with geology, or economic geography with economics), most geographers have viewed such moves with concern. They have thought the distinction between natural and manmade phenomena unhelpful since it obscured some of the essentially integrating characteristics of geographical study, and therefore undermined one part of its long-term rationale as a university discipline.

A GEOGRAPHIC TRINITY

What then are these essential geographical characteristics and why were they considered so important? I think that at least three can be readily identified.

The first characteristic is an emphasis on location. Geography is concerned with the locational or spatial variation in both physical and human phenomena at the earth's surface. It tries to establish locations accurately, to represent them effectively and economically through maps and to disentangle the factors that lead to particular spatial patterns. In human geography it may also propose alternative spatial patterns which are more equitable or more efficient. It is significant that many of the techniques developed within geography from the study of such spatial variation are general in character and not specific to natural or social phenomena.

A second characteristic is geography's emphasis on land and people relations. Here in this essentially ecological approach the stress is on the links between aspects of the natural environment of a particular area and the human population occupying or modifying it. In this type of analysis geographers shift their emphasis from spatial variations between areas (these may be thought of as horizontal bonds) to man–environment links (vertical bonds) within a bounded geographical area. It is worth noting that the bonds may be two-way (e.g. the impact of people on land, as well as land on people), and that the area of interest may be anything from the globe itself to a very small locality.

A third characteristic of geography is regional synthesis, in which the spatial and ecological approaches described above are fused. Appropriate spatial segments of the earth's surface, usually termed regions, are identified, their internal morphology and ecological linkages traced, and their external relations established.

This threefold character of geography is now widely accepted, although individual geographers or university departments will lay emphasis on one aspect of the trinity at any one time. But debates and differences of approach continue to run through the discipline and we need to take some account of these.

GEOGRAPHY AND SUPPORTING FIELDS

Geography is particularly dependent on the flow of concepts and techniques from other sciences.[18] For example, in regional climatology we adapt models originally developed by meteorologists, who in turn draw their concepts from basic physics. Likewise, our models of regional growth borrow from the econometrician.

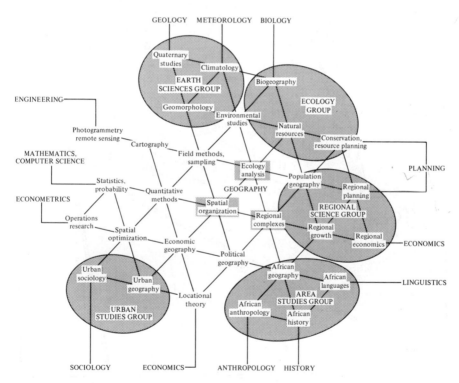

Figure 1.3   Links between geography and supporting fields. African studies (lower right) has been used to illustrate area studies: a similar diagram could be constructed, for example, for Latin American or South Asian studies.

Source: Peter Haggett, *Geography: A Modern Synthesis*, 3rd revised edn. New York: Harper and Row, 1983, fig. 25.8, p. 616.

This dependence underlines the fact that sovereign subjects are about as un-independent as sovereign states. Good botanists need to be reasonable biochemists, good engineers need to be fair mathematicians, and so on. For the student, familiarity with these supporting fields is normally obtained by taking parallel courses in other departments. Geography is unusual (perhaps promiscuous) in the range of its trading partners. A few geographers would argue that special importance attaches to mathematics because it provides a

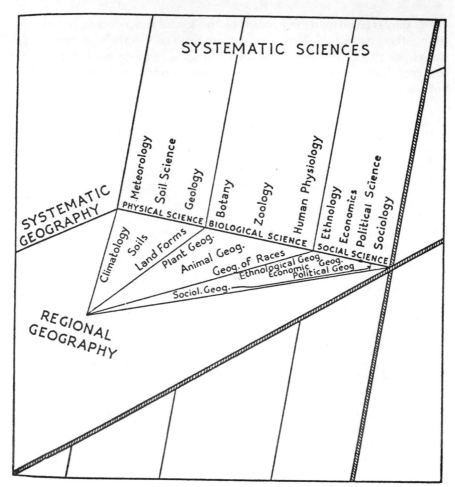

Figure 1.4 The geographic plane. Richard Hartshorne's view of the interrelations of the systematic sciences with geography as two intersecting planes.

Source: Richard Hartshorne, *The Nature of Geography: A Critical Survey of Current Thought in the Light of the Past.* Lancaster, Pa.: Association of American Geographers, 1939, fig. 1, p. 147.

common language in which geographers can express spatial, ecological and regional concepts in a concise and comparable way.

The ways in which geography interacts with supporting fields is shown in figure 1.3. This subject map is important in guiding both student choice and research training. Let us assume that your particular interest is in the humanities and that you are especially attracted to the study of one of the major cultural areas like China. Then the appropriate courses to support regional courses in East Asian geography might include the history and economic structure of China; clearly, courses in Cantonese or Mandarin would be needed for more serious research. By contrast, if you wanted to specialize in environmental problems, you would need earth science courses like hydrology or oceanography, perhaps supported by a course in resource economics.

But this map of geography is too simplistic. As Richard Hartshorne has deftly illustrated, geography and other fields interact in a more complex way.[19] Figure 1.4 shows this as the intersection of two planes. The horizontal plane is represented by geography, the vertical plane by the other systematic fields. These in turn are divided into physical science, biological science and social science. At the intersection of the two planes, the systematic subject calves off a geographical variant. For example, among the sciences meteorology generates a geographical sub-species such as regional climatology, and botany links with plant geography; in the social sciences, economics with economic geography. Even in the humanities, a literary geography has emerged with some geographers finding keys to regional character in the writing of regional novelists like Thomas Hardy or D. H. Lawrence.[20]

But Hartshorne's diagram goes a step further. Within the geographical plane, the separate systematic geographies are focused on a particular region. Regional geography represents a synthesis of parts of systematic geography relevant to a particular region. Notice that not all of the systematics will be taken into account and those that are will be brought in to different degrees and in different sequences. Like the coffee blender, the skill lies in choice. One presidential address to the Association of American Geographers suggested this selection and synthesis made up 'The highest form of the geographer's art'.[21] Certainly it will be different for each region. Hartshorne's regional study of the Silesian basin picked out political history and geographic location as critical elements, while Stoddart showed how understanding the regional problems of Bangladesh demands not only knowledge of the natural hazards of typhoon and flood, but also of demography and diarrhoea.[22]

REGIONAL AND SYSTEMATIC GEOGRAPHY

In his *Geography in Relation to the Social Sciences*, Isaiah Bowman tried to set out how that regional integration worked:

> Geography systematically brings the distributional facts together in their regional framework. If it merely recombined data from other sciences it would be a card catalog, not a science. It goes much further since its main purpose is regional analysis and if possible correlation: the identification of interrelations, the way in which the forces of the environment 'hunt in packs' and produce group effects ... put together [they] form a statement of resources and limiting conditions.[23]

Regional geography represents then a synthesis of those parts of systematic geography relevant to a particular region (see figure 1.5).

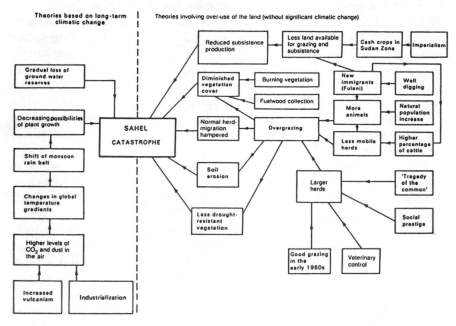

Figure 1.5   Intertwining of multiple factors in a regional problem. The case of the Sahel famine areas in Africa. Climatic change factors are shown on the left, other factors on the right.

Source: Adapted from A. Reenberg, *Det katastroferamte Sahel.* Brenderup: Geografforlaget, 1982; by Arid Holt-Jensen, *Geography: History and Concepts,* 2nd edn. London: Paul Chapman, 1988, fig. 5.8, p. 147.

But how do geographers select what has to go into this regional synthesis? In practice, the selection rule used is regional dominance. Some factors tend to dominate others in a particular regional setting, but may be weak elsewhere. A good rule of thumb is to select as few factors as possible to explain the spatial patterns.

Conventionally, physical elements of the environment were placed first in any regional description because their rate of change was slow, compared say to geopolitical factors which might move overnight. Research has now shown that environmental change can also occur quickly, but has been less important in revising this view than the realization that each technical, cultural or political change may call for a reassessment of those physical resources. The forbidding mountains of one generation become the tourist draw of the next, the tundra of one decade is seen as the oil resource of the next, the strategic island of one year is overflown the next. As Albrecht Penck observed, 'the region represents a state of balance which is immediately disturbed if even a single one of the factors determining it is changed.'[24] So the composition of the light focused on the region in figure 1.4 is constantly shifting.

In regional geography stress is laid not on the spatial dimension of each component alone but on particular associations of environment or population characteristics in particular areas. These in turn give rise to distinctive sub-regions within the general overall area being considered. And so the Russian doll of region hidden within region is revealed.

Where the evidence is available, stress is laid on the stability or instability of such regional structures over time. Which are static and apparently timeless? Which are fragile and ephemeral? And how can we integrate the phenomena within a single area and describe the combinatorial array of links between them? I shall argue later that some form of systems analysis provides one way of attempting to reduce this complexity to a simpler architecture in which it may be better understood.

Systematic studies in geography have a different purpose. They take one (or at most a few) aspects of the human environment or the human population and study their varying performance over a predefined geographical space. Such studies are usually labelled with reference either to the phenomenon concerned or to the sub-field of the natural or social sciences with which it may be identified. Thus a spatial study of voting behaviour may be termed electoral geography (under the 'phenomenon' label) or political geography (under the 'discipline' label). This second type of labelling causes dilemmas in library classification. It commonly leads to student headaches as books on geography are shelved in so many different sections of a library.

ON THE SEARCH FOR SPATIAL ORDER

A critical question in all scientific investigation is just what we are prepared to accept as a signal or a pattern, and what we dismiss as background noise.[25] If we ask of a given world region whether its settlements are arranged in some predictable sequence, or its land-use zones are concentric, or its growth cyclical, then the answer largely depends on what we are prepared to look for and what we accept as *order*. Order and chaos are not part of nature but part of the human mind: in Sigwart's words, 'That there is more order in the world than appears at first sight is not discovered till the order is looked for.'[26] Chorley has drawn attention to Postan's lively illustration of this problem as it afflicted Newton, newly struck on the head by an apple: 'Had he asked himself the obvious question: why did that particular apple choose that unrepeatable instant to fall on that unique head, he might have written the history of the apple. Instead of which he asked himself why apples fell and produced the theory of gravitation. The decision was not the apple's but Newton's.'[27]

We do not have to look far to demonstrate that order depends not on the geometry of the object we see but on the organizational framework in which we place it.[28] Figure 1.6 gives three familiar examples of observational 'flips', which remain well worth repeating. Diagram (A) shows to some an old Parisienne, to others a young woman (after the style of Toulouse-Lautrec). All normal retinas 'take' the same picture; and our sense-datum pictures must be the same, for even if you see the *vielle femme* and I the *jeune fille* the pictures we draw of what we see may turn out to be geometrically indistinguishable. Organization is not itself given by the lines or shapes; were there no preconceived model we would be left with nothing but an unintelligible pattern of lines. Consider the second diagram, (B). There some may see birds (each looking to the left), others may see antelopes (each looking to the right). But would people who had never seen an antelope, but only birds, see antelopes in this drawing?

A third example is given in (C). This pattern of black and white shapes is simply one of melting snow on a grass surface. But it also has a human interpretation:

> The upper margin of the picture cuts the brow, thus the top of the head is not shown. The point of the jaw, clean shaven and brightly illuminated, is just above the geometric centre of the picture. A white mantle . . . covers the right shoulder. The right upper sleeve is exposed as the rather black area at the lower left. The hair and beard are after the manner of a late medieval representation of Christ.[29]

Figure 1.6 Spatial order and meaning. Patterns of lines with alternative (A and B) or hidden (C) interpretations.

Source: Original diagrams in Boring, *American Journal of Psychology*, Vol. 42 (1930), p. 100 and P. B. Porter, *American Journal of Psychology*, Vol. 67 (1954), p. 550. Reproduced from N. R. Hanson, *Patterns of Discovery: An Inquiry into the Conceptual Foundations of Science.* Cambridge: Cambridge University Press, 1958, figs 2, 5, and 7, pp. 11, 13 and 14.

Once the face of Christ has been seen, then we live with it for life; I find it impossible to 'unsee' that pattern and thus may miss other and equally arguable interpretations.

The examples given in these figures will be familiar to many readers. But we often fail to appreciate how much the lessons they teach have

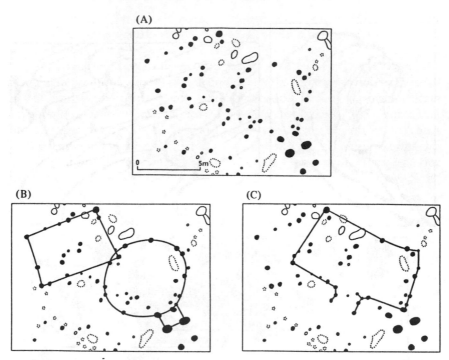

Figure 1.7   Alternative interpretations of a map. (A) Map from an archaeological excavation showing holes marked in black. Some of these may be post-holes for the timbers used in building a hut, others random excavations by burrowing animals. One group of archaeologists selects a set of holes as post holes and suggests a circular hut form (B), while another group favour a larger rectangular hut form (C). Both interpretations bring pre-existing hypotheses to bear on which pieces of the spatial evidence may be regarded as 'signals' and which are irrelevant 'noise'.
Source: Richard H. Gregory and E. H. Gombrich (eds), *Illusion in Nature and Art*. London: Duckworth, 1973, fig. 32, p. 00.

implications for geographic observation. Consider the three maps in figure 1.7. The upper diagram a shows a small 12 m × 16 m section of Salisbury Plain in southern England, with the blackened areas representing depressions in the old soil surface of this chalk downland. The field archaeologists interpret these depressions as the remnants of post holes dug to support the main timbers of a hut. Questions now arise as to the age of the settlement and which tribal groups were the likely builders. To try to answer these, the original map was sent to two archaeologists in different universities for an independent opinion. Maps B and C show the two interpretations. On the left the data have been linked to show a hut with a rectangular structure typical of one period of settlement; the other a circular hut from a more primitive period. Exactly the same data have led to different models; the difference lies not in the evidence but what is seen in the data.

For geography the danger of model imposition is acute. For of all sciences it has traditionally placed greatest emphasis on 'seeing'. In how many field classes have we asked students to 'see' the clues of glaciation in the landscape? The 'seeing eye' beloved of the late S. W. Wooldridge in his study of erosion levels in south-east England is a necessary part of our scientific equipment. But pattern and order exist in knowing what to look for, as well as how to look. And the order of one generation may be the chaos of the next.

A PATTERN OF ESSAYS

Geographers, like archaeologists, are not immune from imposing models on data to make a more consistent tale. My own biases in the rest of this book will be evident. I have placed emphasis on asking questions about the order, specifically the spatial order, shown by geographical phenomena. But readers need to know that this is only one aspect of geographical methodology. Other geographers would have chosen to emphasize other things, to see other patterns.

The organization of the remaining essays in this book is simple. They fall into two groups: 'The Search' (chapters 2 through 5), which reflects on the nature of geography, while 'The Searchers' (chapters 6 through 8) reflects on the nature of geographers.

Thus in section I look at the ways in which geographers have polished and adjusted their mirrors to get particular images of the world around them. Chapter 2 on 'Levels of Resolution' asks how clearly images can be resolved. What can we see? And at what levels of resolution? Chapter 3 on 'The Art of the Mappable' questions the accuracy of the reflection. What causes it to be distorted and blemished? Can we do anything to correct this? Chapter 4 on 'The Regional Synthesis' asks questions about how the glass is focused. Do we try to look at the whole globe, or merely at some limited parts of it? And if the latter, then how are these parts to be selected? And then I explore how these findings are passed on to our colleagues in other fields. Finally in chapter 5 on 'The Arrows of Space' I go back to Barbara Tuchman and Freeman Dyson to explore the time dimension. How do we capture images that are shifting over time? And can we use them to make guesses about the future world?

In the last three essays, entitled 'The Searchers', I ask geographers to be narcissistic and to look into the mirror at themselves. What sort of image do they see there? What questions does it prompt about the discipline we follow and the profession we represent? How well do we do our jobs? How do we relate to colleagues in other fields of enquiry? Is geography now too important to leave to geographers?

Like the spirits haunting Ebenezer Scrooge in Charles Dickens's *A Christmas Carol*,[30] there are three visitations. In the first we look at the spirit of geographers past (chapter 6 on 'Family History'), in the second at geography present (chapter 7 on 'Shifting Styles'), and in the third at geography to come (chapter 8 on 'Geography Future'). The thread running through all three chapters is the question of geographic understanding. How have ideas evolved and been transmitted in the past? How do we decide whether we are making progress? What are our opportunities for further insights in the future?

Although the historical record is a patchy one, with progress and regression intermixed, ideas won and lost, the book has an optimistic conclusion. I argue that if we fail to make gains in the future the fault will be ours alone. Geographical problems stand clamouring to be tackled; resources are available on a scale undreamt of by our founding fathers. The field is wide open and we need bright and dedicated young people to stream into it.

PART I

The Search

Geography Through the Looking Glass

2

Levels of Resolution

All this time the Guard was looking at her, first through a telescope, then through a microscope, and then through an opera-glass.
Lewis Carroll, *Through the Looking Glass and What Alice Found There* (1872)

Two of the greatest intellectual achievements of twentieth-century science are concerned with spatial structures. The first, the general theory of relativity, describes the forces of gravity and the way it shapes the large-scale structure of the universe; that is from only a few miles out to the size of the observable universe (perhaps one million million million million miles). The second, quantum mechanics, deals with strong forces working on small-scale sub-atomic structures at sizes such as one-millionth of one-millionth of an inch.[1]

Equidistant from the stars and the atoms lie the familiar middle-ground distances of the geographer: here are spatial structures which come half-way between the worlds of Einstein and Bohr. But even though the window of geographical interest is a small part of this greatest outer and inner universe, it too has its immense scale variations and one of the fixations of geographers is with the level of resolution needed at each scale.[2]

There are several ways of showing this. One of my favourites is figure 2.1 which shows this scale as a continuum running from the reality of the electron microscope, up to the one-to-one reality of our everyday life, through to the astro-physical reality of the galaxies and on finally to the dimensionless reality of mathematics. Geographical research then occupies a rather well-defined position in this succession.[3] It ranges from highly localized studies (say, of a small atoll, an individual settlement or a small river basin) at magnitudes of 10^{-1} square miles through to worldwide studies of the order of 10^7 square miles. Unlike the microscopic sciences, where small features have to be brought up by magnification to a scale where our eyes and our minds can understand them, geography is macroscopic in that it has to shrink very large features to make them comprehensible.

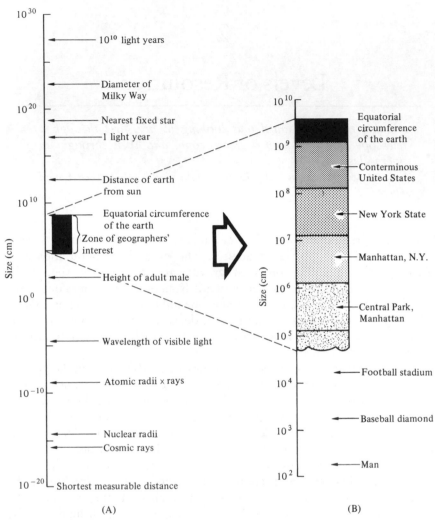

Figure 2.1 Geographers' window of interest within the scale continuum. This stretches from galactic space to sub-atomic space. The zone of interest to geographers is expanded in the right-hand diagram.

Source: Peter Haggett, *Geography: A Modern Synthesis*, 3rd revised edn. New York: Harper and Row, 1983, fig. 1.11, p. 19.

MACROS AND MICROS

Let me illustrate this notion of scale by taking two geographical studies on the same topic: locating the origins of world agriculture. One is at the global scale, which we will term macrogeography; the other at the back-garden scale, which we shall call microgeography. I shall try to show that each is essential to the other.

The origin and location of the world's agricultural hearths (i.e. starting places) have been the subject of intense academic debate. Archaeological evidence indicates the successful domestication of plants and animals by 8000 BC in the Middle East, in the hills of what are now Iraq and Iran. Other finds reveal similar activity in scattered spots in India, northern China and central Mexico. Just where in the world did agriculture begin?

One of the most controversial answers to the question has been provided by the Berkeley geographer, Carl Sauer, in his Bowman Lectures at Columbia University in 1952.[4] His ideas are set out in figure 2.2. Sauer suggested tropical southern Asia as the main Old World centre for agricultural systems based on reproduction by vegetative planting (i.e. subdividing an existing plant into several parts, each of which grows into a new plant), with subtropical Asia Minor as a secondary and later centre for agriculture based on reproduction by planting seeds. This fundamental distinction between the two types of plant propagation is repeated in the two further centres (again, tropical and subtropical) suggested for the New World.

Sauer chose his hearth areas on the basis of the overlap of four spatial requirements. Areas had to be rich enough to have a great variety of plants and animals and thus large enough gene pools for experiment and hybridization to occur; to be outside areas of chronic food shortages; to be outside the large river valleys where advanced water control was necessary for agriculture; and, finally, to be in wooded rather than grassland country. Only small pockets of the humid tropics, where rivers provided regular food supply and allowed sedentary settlement, met all four of Sauer's requirements.

This global thesis was based not on the patchy archaeological evidence, at that time especially fragmentary in the tropics, but on a logical framework of environmental argument. By its sweep and provocation, it stung many other scholars into looking again at the evidence available in their own smaller but more carefully researched areas.

Typical of this microgeographic response was the work of the botanist Edgar Anderson. In his essay on 'Rubbish piles and the origins of agriculture' he examines how Sauer's ideas fit into his own plant studies:

> When I first went to live in San Pedro Tlaquepaque, a small pottery-making town in western Mexico, I was under the mistaken impression that my Mexican neighbours had nothing but rubbish heaps and a few trees in the yards behind their homes. As I lived there longer and came to know about the life of the village, I realized that on any of these heaps were carefully managed gardens.[5]

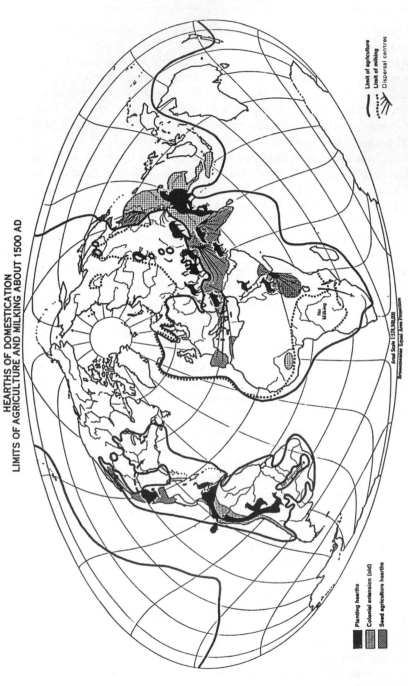

Figure 2.2 Macrogeography. Sauer's views on the global origins of agricultural hearths and the limits of agriculture at about AD 1500. The map projection used is the Breisemeister Equal Area Projection.

Source: Carl O. Sauer, *Agricultural Origins and Dispersals* (The Bowman Memorial Lectures, Series Two). New York: American Geographical Society, 1952, plate IV, facing p. 84.

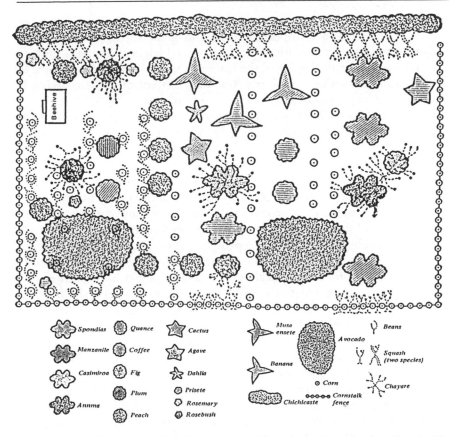

Figure 2.3 Microgeography. Diagrammatic map of an orchard-garden in the Indian village of Santa Lucia, Guatemala (above), with key to plants (below). The scale of the map is not given but is described by Anderson as 'a small affair, about the size of a small city plot in the United States' (p. 120).

Source: Edgar Anderson, *Plants, Man and Life*. London: Andrew Melrose, 1954, figs 13 and 14, pp. 122–3.

In spare time from his main research on maize genetics, Anderson studied more and more of these gardens as he travelled in Central America and got to know the area intimately. Figure 2.3 shows a garden he charted in the Indian village of Santa Lucia, Guatemala. He found it covered by a riotous growth of vegetation so apparently planless that any middle-latitude gardener would have surmised that, even if it had once been a garden, it must have been long deserted. Yet, when carefully charted with the Indian owner's help, Anderson found a clear spatial order. Moreover, noxious weeds were entirely absent, no plant was not of economic value, and efficiency of return in terms of man-hours of effort was high. There were fruit trees, native and European, growing in great variety, a few coffee bushes, giant cacti grown for their fruit, two varieties of maize and of bananas and over it all clambered the luxuriant vines of various cucurbits. It combined the European and North American concept of garden, orchard, medicinal garden, compost heap and bee-yard all in one.

The link to Sauer's work is that Anderson found his hit-or-miss tropical gardens fitted perfectly with Sauer's theory that agriculture may have originated among a sedentary fisherfolk.[6] He sees the notion of refuse-heaps as of extraordinary interest in connection with the origin of cultivated plants because they were open habitats, always being disturbed and always receiving new stock where mongrel varieties of plants could grow up.

This interplay of scales in geographical work is reminiscent of a musical structure. Grand themes are taken up on one level of symphonic construction by the whole orchestra, to be repeated in small passages on particular instruments; a toccata and fugue of research at different geographical scales.

These familiar problems of scale and resolution must have been in my mind when I bought my first atlas in 1945 at the cost of twelve shillings. In addition to the regular world maps it had a black and white section of maps showing the various World War II campaigns where I could follow the fortunes and misfortunes of those family members and village friends who had been called up for military service. The flamboyant signature I'd written in purple ink inside the cover was followed by an equally pretentious address which gave a sequence running through the familiar litany of house, road, village, postal town and county. But then it continued on out with the country, the continent, the planet, the solar system, the galaxy and space itself. It was clear that if I had lost the atlas anywhere in the universe outside a black hole the finder would have known exactly to where it should be returned!

WORLD EXPLORATION

The problem posed by any subject which aims to be global is simple and immediate: the earth's surface is so staggeringly large so that, even if we omit the sea-covered areas, each geographer on a recent international list (there are some 10,000 of them) would have an area of some 6,000 square miles to cultivate.[7] Of course, this kind of calculation is ridiculous and I know of only one country where geographers once worked in that way. But if we follow the argument in the opening essay that one purpose of geography is to describe and interpret the variable character of the earth surface (what Sauer called a 'focused curiosity'), then we need to be aware of the magnitude of the task which is set.[8]

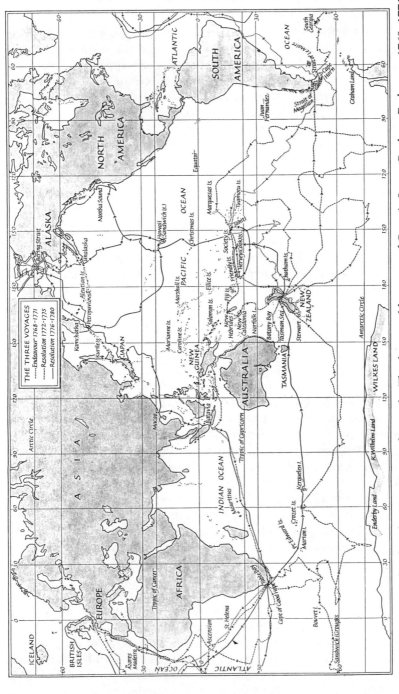

Figure 2.4 Geographical exploration at one of its peaks. The three world voyages of Captain James Cook on Endeavour 1768–1771 and Resolution 1772–1775 and 1776–1780 which accurately filled in large spaces on the Pacific and Antarctic map.

Source: J. C. Beaglehole, *The Journal of Captain James Cook.* London: The Hakluyt Society, 1974, map 5, facing p. 704.

This can hardly be regarded as a new problem for geographers. From at least the time when Eratosthenes roughly estimated the earth's magnitude (*c*.200 BC), the sheer size of the problem has been apparent. Before the satellite age with its familiar earth views, our predecessors were more keenly aware of its importance. Many a doubtful isopleth is now swept self-importantly across areas that our more honest forbears might have filled with heraldic doodles or labelled 'terra incognita'. For most of the history of geography the classic response to finding out about the earth was exploration and discovery. Let us take just one example.

For me, no individual better illustrates the exploration tradition in geography than James Cook.[9] Born in Yorkshire in 1728, he was apprenticed to a North Sea shipowner and volunteered for naval service in 1755. In the war with France he charted the St Lawrence and the Newfoundland coast. His reputation as a navigator, hydrographer and seaman rose rapidly and his calculations and observations were published in the *Philosophical Transactions* of the Royal Society. At the age of forty he was commissioned by the Admiralty to command HMS *Endeavour* in the exploration of the South Pacific.

Geographers had long speculated that there must be a massive southern continent somewhere in that hemisphere to 'balance' the land masses in the northern hemisphere. On this voyage and two which followed, Cook added more to our knowledge of the geographical problems of the southern oceans than all his predecessors. As figure 2.4 shows, the three voyages zigzagged over great areas of the Pacific. Island chains were demarcated, the east coast of Australia and the west coast of North America mapped, and the myth of the southern continent exploded (although Cook rightly conjectured the existence of Antarctica). Typical of his charting was that the North and South Islands of New Zealand were circumnavigated and mapped so accurately that only two significant blemishes can be recognized: the Banks peninsula is shown as an island, while vice versa Stewart Island is shown as a peninsula (plate 3). Cook's accomplishment stands unequalled both in the magnitude of his mapping work accomplished in so short a time and for the accuracy of that work. His biographer said that, while there were innumerable statues and inscriptions, '. . . Geography and Navigation are his memorials'.[10]

John Elliott, who sailed in the *Resolution* in 1772 at the age of fourteen, made his own records of the voyage, including a chart of the ship's track. Like all records it was impounded at the end of the voyage but, like an undergraduate asking for his marked essay, Elliott asked for his promised document back and was invited by Cook to breakfast with him. His chart was returned inscribed in Cook's own hand *Elliott's Chart and Ship Track*.[11] It remained the boy's venerated reminder of sailing with an explorer who added more to knowledge of the world than any before or since.

It is on such fragmentary evidence, sketchmaps and fieldbooks, logs and travel diaries, bearings and distances, that a rudimentary knowledge of the vast spaces of the world was made up. Not until the first satellite photographs did the true vastness of the revolving planet come firmly into focus.

While something of that exploring tradition remains, contemporary geography may have lost too much contact with these roots. David Stoddart recalls the lament of the much-travelled novelist, Joseph Conrad, who feared many decades ago that geography had come to be controlled by

> ... persons of no romantic sense for the real, ignorant of the great possibilities of active life; with no desire for struggle, no notion of the wide spaces of the world – mere bored professors ... their geography very much like themselves, a bloodless thing with dry skin covering a repulsive armature of uninteresting bones.[12]

But the choice is not between the traveller-explorer of an early century and a computer-tied geographer of the late twentieth century, but rather how we bridge the gap between local field observation and scientific generalizations that hold over a wider spatial domain. It is to that bridge that we now turn.

CASAS GRANDES E SENZALAS

Because geographers could not bring the whole world back into their laboratories, they traditionally went out to conduct field work in small areas. But then a second dilemma arose. How could they see the link between their own local fieldwork and the standard regional courses on the continents of the world which formed their bread-and-butter teaching?[13] I first encountered this scale linkage problem in Brazil when trying to compare factors influencing the spatial distribution of forests in a small survey area in the Taubaté valley (about 100 square kilometres) with a later survey of the factors operating over a three-state area some hundreds of times larger.[14]

South-eastern Brazil was chosen because it provided a classic case of a land-use cycle: from primary tropical forest, through crop and pasture and then back to secondary forest. The driving force for the change was coffee (*Coffea arabica* L.) which was first grown in the early nineteenth century in the coastal region around the then capital city, Rio de Janeiro. In the mid-century the financial returns on coffee beans became so high that plantations were established further and further away from the coast. The plantations burst up the Paraíba Valley towards Saõ Paulo in a twenty-mile wide swath of forest felling and burning. Within another generation it had spread out to the north-west beyond Saõ Paulo and with astonishing rapidity the plantation tract along the Paraíba river collapsed. The coffee groves were abandoned to weeds and cattle and the *casas grandes* (plantation houses) and the *senzalas* (slave quarters) decayed and became overgrown.

The impact of this land-use cycle on the fragile soils of the sub-tropics had been noted by Richard Burton in June 1867:

> Hereabouts the once luxuriant valley is cleaned out for coffee, and must be treated with cotton and the plough. The sluice-like rains following the annual fires have swept away the carboniferous humus from the cleared round hill tops into the narrow swampy bottoms, which are too cold for cultivation; every stream is a sewer of liquid manure, coursing to the Atlantic; and the superficial soil is that of a brickfield.[15]

Ninety years after Burton I found myself tramping across the same landscape and wondering how and where best to measure the soil loss and sediment yield that had resulted from the short-lived coffee epoch. At 150 miles long and 50 miles broad, the Paraíba Valley itself was too large for any detailed study. With local advice and help, partly guided by the availability of air photographs and detailed maps, partly by a convenient jeep based at Taubaté, I selected a small right-bank tributary, the Fortaleza basin. This at around 80 square miles was of more manageable size for a month-long study but, here again, scale considerations entered in.

In 1850 the whole area was covered with forest. About one-third of the basin had been cleared for coffee planting and the rest of the forest cut for timber, charcoal and grazing. By dividing up the main river basin into its many small tributaries, a family of well-defined localities was established, from which twelve were eventually drawn (see figure 2.5). Half of these had been cleared for coffee-growing (the test areas) and half had not (the control areas) (see plate 4).

But here another complication arose. Direct comparison of sediment loss in unrelated sets of plantation and uncleared watersheds would be invalidated by all the factors (other than land-use history) which affect sediment yield performance. Matched pairs of basins had therefore to be selected which were as much alike as possible with respect to the micaschist bedrock, terrain and other erosion-related variables. So at this stage I had to fall back on the familiar principles of 'paired comparisons' which form such an

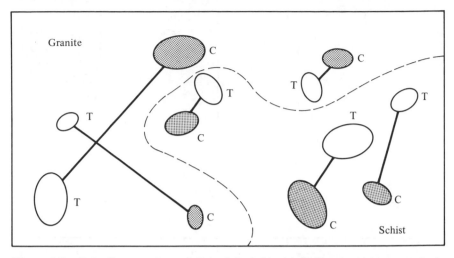

Figure 2.5 Paired comparisons I. Principles of 'pairing' illustrated by watersheds which have been cleared for coffee plantations (T=Test watershed) and left under forest (C=Control watershed). Watersheds are selected to be as alike as possible on factors other than land-use history. Here bedrock and basin size are standardized. The method was used in the author's study of the Paraíba basin, Taubaté county, southeast Brazil.

Source: P. Haggett, 'Land use and sediment yield in an old plantation tract of the Serra do Mar, Brazil'. *Geographical Journal*, Vol. 127 (1961), pp. 50–9; especially fig. 1, p. 51.

important part of work in genetics. The location of the twinned basins is shown in figure 2.5.

SMALL-SCALE SAMPLING

But the problem was still not solved. Altogether the twelve basins covered nearly 500 acres and the final stage was the location of soil sampling points and pits within each basin. Here a more formal sampling procedure could be invoked.

Sample studies had long been used in geographical research. Robert Platt was acutely aware of the 'old and stubborn dilemma of trying to compre-hend large regions while seeing at once only a small area', and he skilfully used sample field studies to build up an outstandingly clear series of pictures of the varied regions of Latin America.[16] Other geographers have used a worldwide selection of sample studies as the basis for teaching manuals in regional or economic geography.[17]

But I see an important difference between these attempts to use sampling

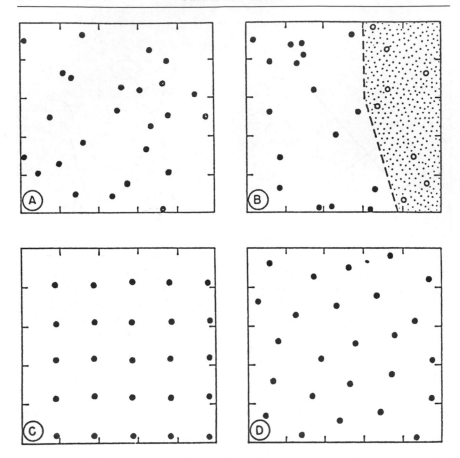

Figure 2.6 Alternative spatial sampling designs. (A) Random, (B) Stratified
random, (C) Systematic and (D) Stratified systematic unaligned.
Source: P. Haggett, 'Scale components in geographical problems'. In R. J. Chorley and P.
Haggett (eds), *Frontiers in Geographical Teaching: The Madingley Lectures for 1963*. London:
Methuen, 1965, fig. 9.2, p. 167.

to circumvent the scale problem and the way in which sampling is now
being used in research. This essential difference is between how samples are
chosen. Sampling theory lies outside the scope of this essay, but it is
important to note that from Ronald Fisher's work in probability sampling
in the 1920s it is possible to estimate how accurate a survey is likely to be
from the information actually collected during the sample survey.[18] This
means that, given a limited budget, the accuracy of any sample survey can
be determined; or, vice versa, given a fixed limit of accuracy, the necessary
size of sample and time-cost estimates can be made. The simplest case of this

type of relationship is in the basic random design where accuracy (the random sampling error) is proportional to the square root of effort (the number of observations).

In practice, geographical sampling poses quite critical design problems. Just to give a taste of these, four commonly-encountered designs are shown in figure 2.6. Where the problem is exploratory and little is known about the characteristics of the geographical 'population' being studied, then a simple random design may be adopted, *A*. Where more is known, the sampling intensity may vary. Thus in a study of land use Walter Wood introduced stratification, *B*, into the random design to allow certain parts of his study area, eastern Wisconsin, to be more heavily sampled than others.[19] The disadvantages of such random designs are that they are more difficult to use as control points in mapping the results and that systematic samples, *C*, may be substituted in their place where mapping is a prime consideration. In an extensive study of the application of sampling design to land-use surveys of flood plains, another geographer, Brian Berry, found that a compromise design, the 'stratified systematic unaligned sample', *D*, gave the most accurate results with the additional advantages of facilitating computer storage and mapping.[20]

Of course, points are only one of the spatial frames. A number of researchers have found line transect methods more accurate than point samples, and there is a large literature on the comparative advantages of the different types of ecological sampling.[21]

Enough has been said to show that the transition from a broadly conceived research problem to actual measurements at specific points on the earth's surface is a multi-stage and messy one. In my Brazilian case, the first two steps, the selection of south-east Brazil and the Forteleza basin, followed a mixed strategy in which I must admit that interest, logistical convenience and the edge of map sheets all played some part. From then on the selection process was more respectable, with both paired comparisons and stratified random sampling having some claim to classical mathematical research design. How much the measures of soil loss and sediment movement that were obtained reflect the selection process still gives me concern. But research designs imposed by severe constraints of time and resources are normal in fieldwork in isolated areas.

LEUKAEMIA AND TWINNED LOCATIONS

While paired comparisons may be used to work downwards from a larger area to a smaller, we may also wish to reverse the process and work upwards from individuals to broader spatial considerations. Membership of

(A)

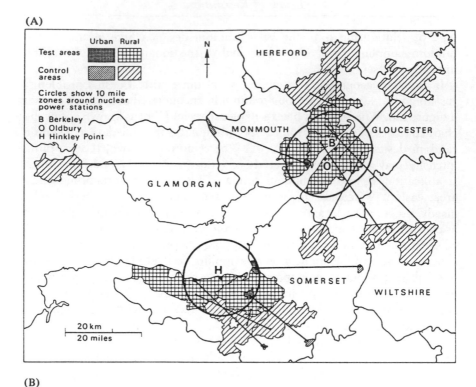

(B)

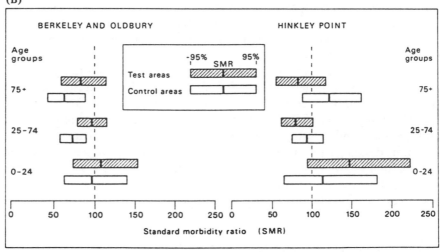

Figure 2.7 Paired comparisons II. Test and control areas for investigating the hypothesis that nuclear power establishments might be related to raised leukaemia rates in children. (A) Location of test and control areas in the Severnside region. (B) Standardized morbidity rates for test and control areas for the Berkeley/Oldbury and Hinkley Point stations.

Source: A. D. Cliff and P. Haggett, *Atlas of Disease Distributions*. Oxford: Basil Blackwell, 1988, plate 3.8, p. 118.

the UK's main radiation watchdog (the National Radiological Protection Board) has inevitably made me study the complex geography of radiation hazards. Public sensitivity in the post-Chernobyl world to the risk, however remote, to the population living near nuclear installations is both understandable and exceptionally acute. This concern includes not only risk from major accidents but what are seen by some commentators as longer-term radiation risks of one of the cancers (leukaemia), particularly among children.

Leukaemia is one of the cancers that the Hiroshima–Nagasaki evidence suggests is especially sensitive to radiation effects. Several studies have appeared to show a concentration of young leukaemia patients into spatial 'clusters', and at least some of these clusters lie in the vicinity of nuclear installations; reports of a cluster of childhood cancers in Lydney across the river from the Berkeley and Oldbury stations on the Severn estuary in southwest England are a case in point. Against this, the large number of statistical tests undertaken provide little support for the notion of space-time clustering, let alone a 'nuclear power station effect' (plate 5).

Given the degree of public concern, governments have set up several studies to try to resolve the problems involved when numbers of leukaemia cases in any one area are small and when results are open to a wide range of interpretations. In the Cook-Mozaffari study,[22] the incidence of, and mortality from, different types of cancer among people living near nuclear installations in England and Wales were compared with the average rates for the regions in which the installations were located and with the levels observed in control areas. To illustrate the method, the Severnside area has been chosen (see figure 2.7).

Control areas outside the 10-mile rings to match against each of the test areas were selected by searching for an area of approximately similar population size located within either the same county or a contiguous county, and of matching rural/urban status; that is, rural district (RD) with RD, and metropolitan borough (MB) or urban district (UD) with MB or UD. In figure 2.7(A), different shading types have been used to discriminate between test and control areas, while the matched pairs have been linked by vectors. Initially it was planned to use a fourth criterion (similarity of social class structure within the populations). This was dropped when it was found that, once the first three requirements had been met, there was rarely a choice of control areas.

As far as possible, controls were chosen within the same cancer registration region. This was important since standards in completeness of registration differ between regions and, until recently, there was no system for identifying duplicate registrations which might contaminate results. Finally, controls were chosen wherever possible for within the same standard region.

Over the twenty-year period of the study, the number of leukaemia cases registered in the Berkeley-Oldbury test area was 185 and in the Hinkley test area 117. The comparative figures for the two control areas were 149 and 138. This overall total of 589 cases in a population of just under one million in the 1971 census has little meaning until it is converted into a comparable form.[23] Thus, figure 2.7(B) shows standardized mortality rates for the test and control areas of both Berkeley-Oldbury (left) and Hinkley (right). Note that the population is divided into three age strata for the young, middle-aged and elderly. Tabulations are by persons rather than for male and female separately.

As a guide to the significance of the geographical variations between the ratios, confidence limits have been derived assuming that the observed values follow reasonable statistical distributions. Here again the geographical circle cannot be squared. Routine tests cannot usefully be made of the differences between cancer ratios in the test areas and those in the matching control areas; the numbers of observations are so small that they are subject to large sampling variability. But analysis of national figures for leukaemia and for all malignancies (involving therefore much larger numbers) shows no consistent pattern of raised incidence in areas adjacent to nuclear installations. Nonetheless public concern remains so high that it would be surprising if detailed study were not to continue and that analytical field studies will be conducted wherever there is a risk of a positive association being detected. Stanley Openshaw's approach, following an original and very different route, may have more success, but I suspect the small differences will prove too elusive to clinch the argument.[24]

REGIONAL TRENDS AND LOCAL RESIDUALS

Quantitative attempts to isolate and measure scale components in geographical patterns are not simple. Perhaps the simplest way into a complex field is to go back to a method I used more than thirty years ago in studying a forest distribution in central Portugal. The basic ideas of the mapping method can be seen from figure 2.8 which shows the stages by which a given distribution may be broken down into what are termed 'regional' and 'local' components. Map *A* shows the original distribution of cork oak (*Quercus suber*) in a section of central Portugal. By covering the area with a rectangular grid the proportion of land under this species in each square cell can be counted. The resulting map, *B*, describes the area in two-dimensional form. Like contour values for terrain it could be converted to a three-dimensional plaster model but in any case we can think of it conceptually as a three-dimensional 'trend surface'.[25]

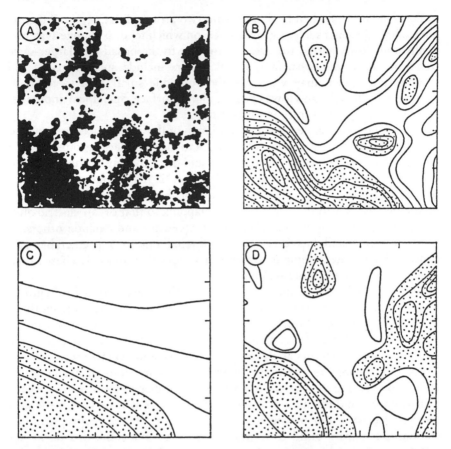

Figure 2.8 Regional and local components in a land-use pattern. Use of filter mapping to separate regional and local components in the distribution of cork oak (*Quercus suber*) forests in the Tagus-Sado basin, central Portugal. The original distribution (A) is shown as a contour surface in map (B). The regional trends are filtered out in map (C) to leave a map of local residuals (D).

Source: Peter Haggett, *Locational Analysis in Human Geography*. London: Edward Arnold, 1965, fig. 9.5, p. 173.

The height of the surface (i.e. a degree of forest cover) at any one point may be regarded as a response to the operation of that complex of geology, topography, climatic peculiarities, natural composition, economic disparities, and the local and regional history which together are thought to determine forest distribution. Variation in the form of the surface may be regarded as responses to corresponding spatial areal variations in the strength and balance of controlling factors.

These factors may then be divided into two groups. First, regional factors controlling elements such as growing season which tend to change systematically and slowly across the area, giving rise to the broad and larger-scale trends in the response surface. Local factors include such items as soil composition, which may vary sharply from one small area to another, leading to local highs and lows in the response surface. Such local anomalies are unsystematic and spotty in distribution.

Map C shows a simple regional trend map of the area. It was constructed simply by drawing a circle around each cell to include a large ground area (it happened to be 2,500 square kilometres) and then calculating the woodland fraction within this circular unit. Plotting this '2,500' surface caused local detail to be lost but the main lineaments of the pattern show up clearly. Nettleton has likened the effect of such mapping to that of 'an electric filter which will pass components of certain frequencies and exclude others.'[26] Certainly the detail has been lost or filtered out in a predictable and controllable manner. Comparison with maps of other areas also based on a similar 'grid' is made more reliable.

Separation of local anomalies can be simply derived from the regional map. For each cell the values of the original cells are subtracted from those of the '2,500' surface. Positive values (i.e. where local values exceed regional values) are shaded and negative values are unshaded. Map D shows the operation of local factors as a pattern of positive and negative residuals.

But trend maps are essentially quantitative expressions of a qualitative choice. By varying the circle size an infinite number of trend-surface maps can be drawn and the nature of the resultant maps will vary with the grid interval chosen.[27] But by including details of the choice geographers can standardize maps around conventional levels to overcome this disadvantage.

THE RIDDLE OF THE SANDS

A second approach to the problem of local and regional variation that cuts out the need for complete information on all the area considered is to go back to sampling.[28] I find one form of sampling, nested sampling, particularly valuable in exploratory studies where there is a need to cover as large a region as possible, but at the same time pay attention to local variations. The basic idea of the nested approach is given in figure 2.9.[29] The Breckland is a distinctive region of sandy deposits overlying the chalk escarpment in eastern England. The origin of the sandy deposits is obscure, but it is commonly believed that they originated from glacial till derived from the Sandringham Sands to the north. The Breckland covers an area of twenty by

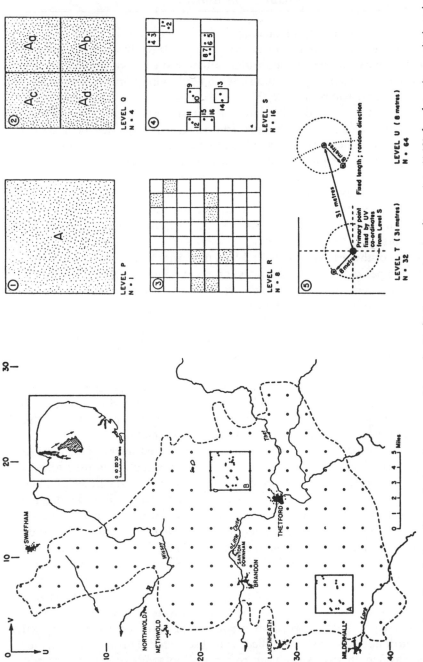

Figure 2.9 Sampling designs for variations by geographic scale. (Left) Systematic sampling points (149) for detecting variations in grain size of surface sands in the Breckland area, eastern England. Sampling points are arranged on a regular grid. (Right) Nested sampling designs at five levels within the two square sampling frames located at A and B on the left-hand map.

Source: R. J. Chorley, D. R. Stoddart, P. Haggett and H. O. Slaymaker, 'Regional and local components in the areal distribution of surface sand facies in the Breckland, eastern England'. *Journal of Sedimentary Petrology*, Vol. 36 (1966), pp. 209–20; figs 1 and 3, pp. 210 and 214.

fifteen miles, now largely planted with conifers with only a few areas of open heathland remaining. On a day in May 1963 it was the scene of unusual activity as eighty Cambridge undergraduates swarmed over the sands armed with soil augers (and sandwiches) taking borings from the top half-metre of the soil profile.

Seen from the air the distribution of the students in their bright anoraks would have seemed puzzling. But in terms of figure 2.9 the logic behind their distribution was clear: they formed part of a carefully thought-out nested sampling design. The purpose of the design was to determine whether there were significant geographical variations in grain size of these surface sands, and if so to what extent it was possible to disentangle the regional and local controls in the spatial distribution. Finally, we wished to find out which spatial scales were associated with the greatest variations in grain size components.

Analysis of the results was at six different spatial levels and showed three contrasting peaks. Regional contrasts (at a wavelength of around seventeen kilometres) made up over half the variation. It reflected large-scale variations in the composition of the original glacial deposits from which the sands were derived. This showed a steady coarsening of grain size towards the north-east, the presumed centre for their wind-blown origin. At the other extreme of the scale were local contrasts with a peak in variability at around eight metres. This made up a further quarter of the variability in sand-grain sizes and was clearly related to the dramatic periglacial sorting processes which produce the cellular polygons and stripe patterns so strikingly present in the Breckland. This local striping pattern was also reflected in the alternating pattern of heathland vegetation.

Much of the remaining variation was concentrated in the range of 125 metres to one kilometre, and features at this wavelength continue to be a puzzle. No features of this size were previously known in the Breckland and none was directly recognizable from air photographs. But by digging through the journals in the Cambridge University Library we came across a paper where periodic features of about this wavelength occurred on sandy deposits in the Netherlands around Barneveld. Here they were thought to be old sand dunes due to Pleistocene wind action. So for the Breckland, with a similar geological history, the most likely explanation is that these factors reflect the degraded remnants of dune fields from a formerly very dry epoch.

By using scale-related sampling the students were able to meet all three original research goals. Two known sources of variation were confirmed and a third, not previously identified, was discovered. Further, periodicities at this wavelength have since been recognized in sands of similar age and provenance in the Netherlands.

MIRAGES

The examples used so far have fallen largely within the domain of physical geography. But the problems of scale apply with equal force in human geography. Here, two almost intractable problems are faced. First, much of the data used in human geography (for example, population data) are released for administrative areas rather than for points; secondly, these areas vary wildly in shape both between countries and within countries. This variation had little more than nuisance value when the limit of sophistication was the choropleth map. But, as more refined statistical analysis is carried out, the problem has grown in importance. Weighting for areas can overcome some of the problems, but there remains a range for error and misinterpretation so great that Duncan found it necessary to devote the greater part of his pioneer book on *Statistical Geography* to just such problems in analysing areal data.[30]

Unless great care is taken, such 'mirage effects' are likely to become more common as more refined indices are derived. The late Maurice Kendall, writing as a statistician, has warned that with certain coefficients of geographical association we can get any coefficient we choose by juggling with the collecting boundaries.[31] It is an open question whether, for example, detailed medical maps in which mortality indices are carefully standardized for age and sex should not equally well be standardized for the areas for which they are collected. Certainly we need to be reassured that the apparently unhealthy small pockets of disease are not simply artefacts of the collecting areas.

FROM ICE AGES TO KONDRATIEFFS

So far in this chapter we have been concerned with the size of geographical space. We have shown both that spatial structures are recognizable on different scales and that these can sometimes be linked to the operation of different constructive processes. But the same scale variability operates in time. From everyday experience and from meteorological observation, we know that air temperatures vary regularly both on a day–night basis and on a summer–winter basis; similarly, irregular change occurs with an air mass change or when runs of warm years occur. At still longer timescales we know from geological evidence of the alteration of tropical and glacial climates. This range of fluctuation in temperature is shown in figure 2.10 as a 'variance spectrum', a diagram in which relative variation is shown on the vertical axis against the spectrum of time in years along the horizontal

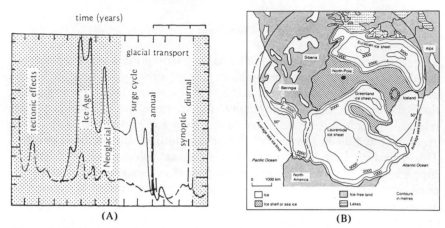

Figure 2.10   Spectra of geophysical events drawn on time and space scales. (A) The broken line shows temperature change and the solid line the relative importance of glacial transport. Both axes are plotted on logarithmic scales. (B) An example of the spatial changes linked to the time spectra. The maximum extent of the Laurentide and Eurasian ice sheets in the northern hemisphere.

Source: (A) M. Church, in R. A. Cullingford, D. A. Davidson and J. Lewin (eds), *Timescales in Geomorphology.* Chichester: John Wiley, 1980, fig. 24, p. 17. (B) David Sugden, *Arctic and Antarctic: A Modern Geographical Synthesis.* Oxford: Basil Blackwell, 1982, fig. 10, p. 100.

axis.[32] Like a radio dial the diagram shows in effect a waveband with long-wave changes on the left, short-wave on the right.

But do such changes relate to space? If we now superimpose on that temperature spectrum the spatial impact of low temperatures (i.e. the area of the earth's surface covered by ice) we can see a different, though partly related, picture. The Ice Ages, with a wavelength of around 100,000 years, make one peak, while glacial surges at around the ten-year wavelength another. Seasonal freezing and melting at the annual wavelength, and even daily changes, form two other peaks at the right hand side of the diagram.

The feature I find fascinating in this diagram is that time is here translated into space and vice versa. The Ice Age peaks translate into a process that covered and uncovered half a continent with glaciers; the annual temperature peak translates into a few metres of ice advance or retreat on the ground. There is a spatial spectrum of change that can be locked into the more familiar time spectra.

Interplay between time and space spectra is not confined to physical processes. We know that economic activity shows similar components over time with recognizable peaks and troughs at particular wavelengths. These are given distinctive names in the literature – Juglar cycles, Kitchin cycles, Kuznetz cycles, Kondratieff cycles and so on. Much of my own work in the

1960s was concerned with trying to map the spatial correspondence between short-term business cycles and their impact on small regions.[33] More recently, attention has turned to the geographical implications of the long Kondratieff waves which have a global impact and may involve the rise and fall of great regimes, such as the recent surge of the Asian Pacific economies.[34]

ALLOMETRY, FRACTALS AND SCALE

In stressing the importance of resolution levels in geographical space I have tried to illustrate simple cases where changes over this spectrum are important. It helps to explain why geographers, in studying the large world, are so sensitive to the levels and boundaries at which they apply. Such sensitivities have been neatly summarized by McCarty:

> In geographic investigation it is apparent that conclusions derived from studies made at one scale should not be expected to apply to problems whose data are expressed at other scales. Every change in scale will bring about the statement of a new problem, and there is no basis for assuming that associations existing at one scale will also exist at another.[35]

But it would be misleading not to pay tribute to those who have worked along other paths and tried to find uniform processes operating over a wide range of scales.

Back in the 1930s the biologist Julian Huxley developed ideas of allometric regularity where biological change at many sizes, from cell to organism, could be interlinked. More recently Benoit Mandelbrot at IBM has explored the scale-free properties of fractal geometry,[36] in which immensely complex structures can be seen to stem from simple recurrent equations. Like Stephen Hawking's unification of physics,[37] the time may come when we can see these structures as all part of a common pattern. But, for geographers at least, that horizon still lies some way off.

3

The Art of the Mappable

For some minutes Alice stood without speaking, looking out in all directions over the country — and a most curious country it was ... 'I declare it's marked out just like a large chessboard!'
Lewis Carroll, *Through the Looking Glass and What Alice Found There* (1872)

As a geographer, it has always seemed appropriate to me that my first appearance in print was on a map (plate 6). January 1952 had brought dramatic floods to the low counties around the southern North Sea. Then a Cambridge undergraduate, I joined the teams from my college who headed north to the Wash in specially chartered charabancs. There we formed ourselves into competing chains to pass sandbags from the nearest dry road up to strengthen the top of the dykes threatened by the next high tide. In the event the worst holes were blocked and an inundation which had killed six people and flooded millions of acres was contained.

But major changes had occurred in the line of offshore bars and islands which formed a natural protective breakwater off the north Norfolk coast. One of these, Scolt Head Island, had been a subject of prolonged study by the Cambridge Geography Department and by Alfred Steers in particular, and had been mapped and remapped by a succession of Cambridge students that went back to Oskar Spate and Ronald Peel in the 1930s.[1] In the summer of 1953 Steers allocated John Small and me to remap Scolt, while Michael Chisholm and Derek Brearley were given Blakeney Point. Our revised maps of the storm damage were constructed largely by levelling and plane-tabling, revising section by section the earlier six inch to the mile maps of the area and thus plotting the effects of the storm. They were later published by the Nature Conservancy Council.

CONVENTIONAL MAPS

Characteristic of this sort of field mapping is that there is a direct and linear relationship between the sections of the earth's surface you are mapping and its formal representation on a two-dimensional sheet. The map is a miniaturized but recognizable portrait of the real world. I can show this relationship most clearly by a specific example. Figure 3.1 shows another of the maps I produced with undergraduate colleagues the following summer in northern Portugal.[2] In this case, the area was the Ancora valley in north-west Portugal, a river with a small catchment draining to the Atlantic near the Spanish border. Our objective was to study land use changes in this rural economy in the period immediately following improved road building. In addition to a sketch map survey for the whole valley adding detail to a standard 1/25,000 topographic sheet (figure 3.1A), five small areas were chosen for special study. Field surveys were made of these using the geographer's usual toolkit – a combination of plane tabling, compass traverse, chaining and levelling. Figure 3.1B shows one of these maps, for the small five-acre farm of Vila Verde in the central part of the valley. This was a relatively prosperous property and had, like most farms in the valley, a complex water supply system for irrigation; the map shows the rich mosaic of crops grown on the terraced fields.

If you compare this map with a ground photograph of the same area (see plate 8) taken at the time of the survey you can see the direct correspondence between the two. If the photograph had been vertical, the correspondence would have been still greater. This counterpoint of map and field, field and map, is at the heart of geographical exploration. As a leading British geographer, S. W. Wooldridge, commented in *The Geographer as Scientist*, 'the ground and not the map . . . is the primary document . . . Field work consists in comparing the map with the actual ground.'[3]

The history of cartography is the attempt to close the gap between the two, not only in local areas where the plan relationship of field and map is self-evident but in the continental and global picture where maps have to be assembled slowly and carefully from fragmentary numerical data on bearings and distances. Until the advent of remote sensing the whole picture could not be seen. The pleasure and relief that the first satellite image I saw – one of the 'heel' of southern Italy – matched the cartographer's reconstruction will be missed by a generation for whom pictures of a whole hemisphere are the routine diet of television.

A DISTORTED WOLFGANGSEE

Twenty-five years after the Portugal maps I was engaged in map making of a different kind.[4] In the early summer of 1980 with my wife and some of the family I stayed in the small town of St Wolfgang in the Saltzkammer-gut region of Austria. The ostensible reason was to take walks in the

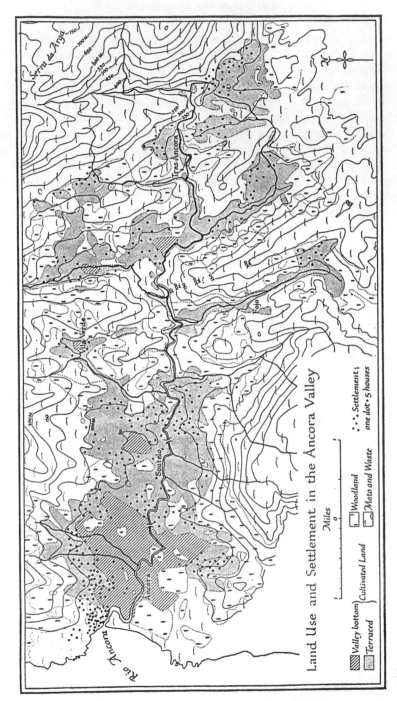

Land Use and Settlement in the Âncora Valley

Miles

| Valley bottom } Cultivated Land | Woodland |
| Terraced | Mato and Waste |

•.• Settlement;
one dot = 5 houses

(A)

surrounding mountains (especially those that had small railways from which you could walk down) and to introduce the children to Salzburg. But one of the problems for a geographer on holiday is that you cannot 'switch off', and it was the lake (the Wolfgangsee) and the little packet boats that crisscrossed between the lakeside villages that caught my interest and forced me to think about the curious spatial structure they reflected.

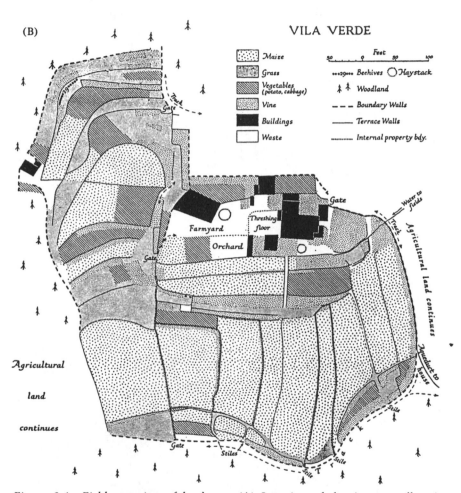

Figure 3.1 Field mapping of land use. (A) Location of the Ancora valley, in north-west Portugal, showing field-survey locations. Note Vila Verde at top centre. (B) Land use on an individual farm, Vila Verde. Compare with photograph in plate 8.

Source: R. D. Hayes, 'A peasant economy in north-west Portugal'. *Geographical Journal*, Vol. 122 (1956), pp. 57 and 63, figs 3 and 7.

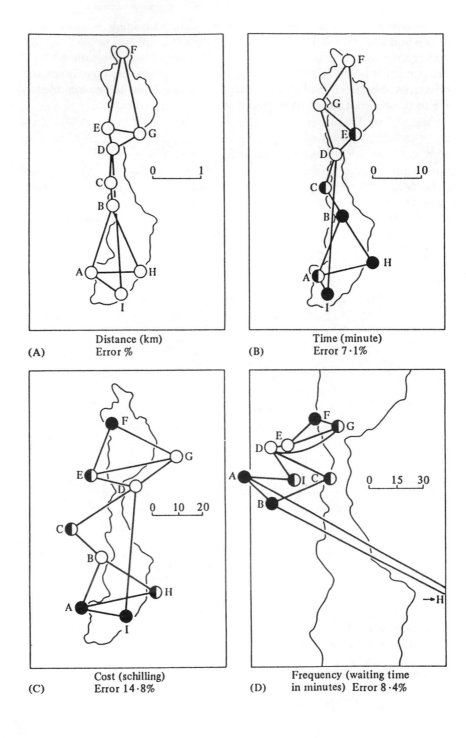

(A)   Distance (km)
Error %

(B)   Time (minute)
Error 7·1%

(C)   Cost (schilling)
Error 14·8%

(D)   Frequency (waiting time
in minutes) Error 8·4%

The conventional map of the Wolfgangsee is shown in figure 3.2A. This is taken from the standard Austrian topographical map for the area and was constructed on the same principles as the Vila Verde map in northern Portugal. But the more I travelled on the boats crossing the lake, the clearer it became that the actual spatial relationship between the lakeside villages was not quite what it seemed. Certainly not that which was shown so accurately on the conventional map. If we take the nine halts served by the state-owned motor vessel service then some (e.g. St Gilgen and St Wolfgang) were on 'fast' routes and well served, while others were on 'slow' routes with many stops and infrequently served. But could all this be put on a map?

Fortunately, the advent of the high-speed computer had enabled the development of some clever and powerful algorithms to convert this kind of information into map form. The methods, given the family name of multi-dimensional scaling or scaling for short, are fully described in the standard texts.[5] The computational procedures are complex but they need not detain us here, except to note three things. First, whereas a conventional map uses essentially distances and bearings, the new scaling maps can accept all sorts of odd data on the spatial relations between places. Second, constructing the map needs billions of calculations; hence the need for the computer. Third, the map will only approximate to the real world and you cannot guess in advance how accurate a map is likely to be. The process is best thought of as the incident in a Tom and Jerry cartoon when a steamroller flattens the cat. The flattened two-dimensional cat is a picture – but a distorted one – of a real 3D cat.

The remainder of figure 3.2 sketches in the results obtained when running the Wolfgangsee through a multi-dimensional mapping program after I got back to Bristol. The transport links formed by the movement of the boats are shown by lines, the nine lakeside halts are shown by circles, and the lake outline is included for completeness. In A the linear distance in kilometres was used and, of course, the map is the same as the conventional map. The journey time in minutes (map B) gives a more complex map with an overall error of seven per cent and the distribution of error between locations shown by the shading; white circles are most accurately shown, black circles

Figure 3.2   Non-linear mapping: I Local. Different spatial configurations for a transport network connecting nine locations around an Austrian lake in the Salzkammergut (the Wolfgangsee). The four different spatial configurations are based on (A) distance, (B) time of journey, (C) cost of journey and (D) frequency of service. Whereas (A) shows a conventional linear mapping, maps (B), (C) and (D) are produced by a nonlinear algorithm. The lake outline is included for comparative purposes in all four maps.
Source: Peter Haggett, 'The edges of space'. In R. J. Bennett (ed.), *European Progress in Spatial Analysis*. London: Pion, 1981, fig. 3.10, p. 68.

are the most inaccurate. When cost (ticket price in Austrian schillings, map D) is mapped the error is doubled and this is reflected by a squarer, less elongated pattern showing that in cost-space the lake is rather circular. But the greatest distortion from the conventional map is shown in map E which is based on the frequency or infrequency of service as measured by the waiting times between boats. This shows that the lakeside villages can be divided into two, a large group of eight well-served places and a ninth place at which the boat calls only twice a day.

But which of these maps is the right one? The answer surely is that each is showing a different aspect of the spatial structure of this settlement. For emergency services, time (map C) may be of the essence; for a stranded hiker, map E may be the more critical. There are many, many maps of each fragment of the earth's surface and they change as the information changes. The last definitive map of any area can never be drawn.

AN IMPLODED PACIFIC

But why is this holiday puzzle relevant? To translate it into something of more general concern, let me take an example that spans a hemisphere – the Pacific Basin. This makes up about one half of the earth's surface and currently the countries around its rim include some of the fastest-growing economies in the world. Conventional maps of areas of this size are difficult to draw since distortion inevitably results in translating a massive segment of the globe on to a flat, two-dimensional piece of paper; squashing half an orange on to a flat surface will give some idea of the distortion involved. The particular map projection used will determine the nature and extent of distortion introduced. Subject to this proviso, all the projections in common use attempt to ensure that either the locations of points on the globe reflect their relative positions on the globe or that areas or directional relationships are preserved.[6] Look back to the two world maps in chapter 2 (figure 2.2 and 2.4) to see two different, but equally valid, world map projections.

Multi-dimensional scaling can be used to convert the Pacific into other metrics. Consider, for example, figure 3.3A. This uses lines joining places of equal time, the technical term being isochrones, to plot the relative time accessibility by scheduled airline carriers of twenty-five places in the Pacific basin; from fast extended 747s crisscrossing the routes between the big cities, to slower local carriers between the small island chains. The placenames associated with the locations used are listed with the map. The isochrones have been standardized so that 100 denotes average accessibility; values less than this indicate superior accessibility and greater values (stippled) demarcate the less accessible parts of the basin. The diagram

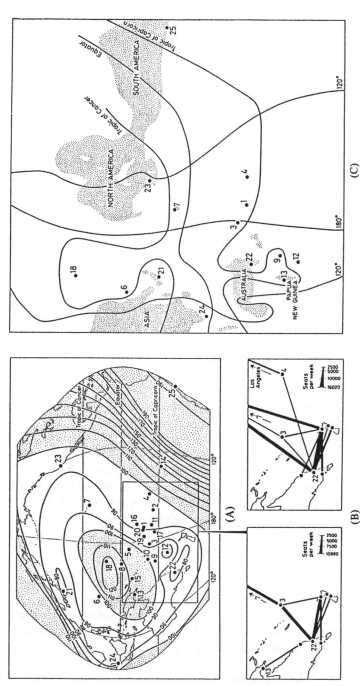

Figure 3.3 Non-linear mapping: II Global. The hemisphere of the Pacific Basin mapped in time–space. (A) Relative time accessibility in 1975 by scheduled airline carriers between 25 centres in the Pacific Basin. (B) Route capacities by international carriers in seats per week in 1975 (left) and 1981 (right) within the southwest Pacific. (C) Data on time accessibility in map A recomputed by multi-dimensional scaling to give a time–space map. The key to places mentioned in the text is (3) Fiji, (4) French Polynesia, (13) Papua New Guinea, (18) Trust Territories of the Pacific, (21) Tokyo, (22) Sydney and (23) San Francisco.

Source: The map is based on unpublished work by Dr P. Forer of the University of Canterbury, Christchurch, New Zealand. The changing air-traffic flows are based on M. Taylor and C. C. Kissling, 'Resource dependence, power networks and the airline systems of the South Pacific'. *Regional Studies*, Vol. 17 (1983), pp. 237–50. Combined maps from A. D. Cliff and P. Haggett, *Atlas of Disease Distributions*. Oxford: Basil Blackwell, 1988, plate 7.2, p. 226.

shows that a large part of the central Pacific centred on the Trust Territories of the Pacific (18) is up to one-fifth more inaccessible than average while, from French Polynesia (4) eastward to the Americas, accessibility falls by a factor of almost two.

The area covered by the maps in B is delimited by the box on A. The maps indicate why the zone of inaccessibility in the central Pacific exists. The left-hand map uses flow lines to show the linkages in seats per week provided by international carriers such as Quantas and Pan Am between various centres in 1975: the right-hand map gives the same information six years later. Not only have route capacities multiplied but the amount of overflying of the Pacific, bypassing local centres, has increased. For example, the maps suggest that Fiji (3) had less stopover traffic in 1981 than in 1975 and so in that sense had become less accessible.

To produce the map shown in C, multi-dimensional scaling has been used to transform the conventional geographical map A into a time metric; the relatively accessible places on map A are now plotted closer to each other on map C, while the relatively inaccessible places have been moved apart. The effect is to push North America and the Far East closer together than they are on a conventional map because of the frequent flights between Japan (Tokyo, 21) and the United States (San Francisco, 23). The inaccessible portion of the central Pacific Basin apparent on map A is now mapped as two outposts. The Trust Territories are moved to the north (note the new position of 18). Papua New Guinea (location 13) is no longer located to the north of Australia but, hernia-like, has burst through that continent to be relocated in the south.

Although the new map is unfamiliar, it shows the locational forces at work in the Pacific in a dramatic way. In the physical world the earth's crust is reshaped by the massive slow forces of plate tectonics. So also technological changes of great speed are grinding and tearing the world map into new shapes. Capturing those spatial shifts is at the heart of modern geography.

THE NIGHTWATCHMEN OF TRONDHEIM

Before leaving non-linear mapping it is worth recalling some of its unusual puzzle-solving properties. My example takes us back to the thirteenth century and demonstrates the utility of the method for handling what the distinguished Cambridge statistician, David Kendall, called 'Odd bits of information'.[7] Consider the following quotation:

Then shall they meet out by Orene and all together go up above Skagen (the headland between Nidelven's mount and the fjord), three shall turn along Geilene and up along the street above Martin's church and from there up west of Gregory's church up along the long street above Kongsgaard to the bishop's boathouse. But the other three turn up along the long street and above the churchyard of Clement's church and up above Kroken and thus up along the wharves and up between Holy Cross church and Soppen and up along the long street between All Saints' church and Benedict's church and up along the broadstreet between Kongsgaard and the house of the Preaching brothers' cloister and up west of Kongsgaard until they meet the others. Thus shall they turn and they go the lower way who before went the upper, in the same way out to Sigurd Svardag's gaard. Later shall they who went the upper way go down along Krupmannastrete and thus along the long street below Olav's church and out by Orene, but the others turn out along the long street above Gregory's church and out by Martin's church and meet at Orene by the smithy. So shall they go every hour until it is daylight.[8]

The text is a translation of a thirteenth-century document describing the routes taken by nightwatchmen as they patrolled the streets of the Norwegian fjord town of Trondheim. The list gives the sequence of churches and roads in the order in which they were visited during the night.

Like most of its Scandinavian neighbours, Trondheim was then a crowded wooden city, and the watchman's job was to keep down the risk of fire while the townfolk were asleep. In a particularly cataclysmic fire in 1681, Trondheim was burnt to the ground and rebuilt on more open and spacious lines which allowed protective firebreaks between the buildings. As a result of this rebuilding, the position of most medieval churches (themselves of wood) and streets was lost. Archaeological excavation had allowed the identification of some church sites, but it was still not clear which medieval church was which.

Patricia Galloway has used multi-dimensional scaling in an ambitious attempt to reconstruct a plan of the medieval city from the nightwatchmen's itineraries.[9] The document cited gives an incomplete 'map' of some seventeen locations as a sequence of streets and churches which are linked or not linked in the itinerary. The best multi-dimensional scaling map is given in figure 3.4 at A. By matching the archaeological and documentary evidence it is possible to anchor a few points on this map on to matching locations in modern Trondheim and thus to reproduce in part the street sequence followed in the long-vanished medieval town. The computer map is hardly accurate enough to tell the archaeologists to 'dig at this point and you will find the missing church', but it does indicate a probable area for its location.

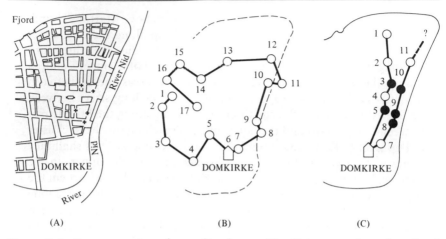

(A)   (B)   (C)

Figure 3.4 Reconstruction of a medieval map. Non-linear mapping of medieval Trondheim, Norway, using a thirteenth-century nightwatchman's itinerary. (A) Modern street pattern with the Domkirke and crosses showing the sites of early-medieval churches as shown by archaeological excavation. (B) Unconstrained best-fit map using non-metric multi-dimensional scaling anchored on the Domkirke.
(C) The best-fit map constrained to align with early medieval church sites.
Source: Patricia Galloway, 'Restoring the maps of medieval Trondheim: a computer-aided investigation of the nightwatchmen's itinerary'. *Journal of Archaeological Science*, Vol. 5 (1978), pp. 153–65.

To sum up. Non-linear mapping provides one example of the ways in which computers allow geographical information to be encoded to exploit the communication properties of the map. It allows a breakaway from the conventional geographical map based on a physical distance separation. It permits distance to be replaced with any other relevant appropriate metric. I have given very simple examples based on cost, time and service and on medieval documents. Such maps are likely to be particularly valuable where they show past or future changes over time. But I hope it will now be clear that the method can incorporate all sorts of information; for example, by

Figure 3.5 Shakespearean space: Romeo and Juliet. The extreme flexibility of non-linear mapping is shown by Peter Gould's map of the play 'Romeo and Juliet' (1610). The input was the number of lines exchanged between the dramatis personae and the maps show (A) the location of the main characters, (B) the contour map of number of lines spoken in the play and (C) a 'regional' division into Capulet and Montague space. Readers will not be surprised that my highest role in school productions was in a peripheral location.
Source: Peter Gould, 'Concerning a geographic education'. In David A. Lanegran and Risa Palm (eds), *An Invitation to Geography*, 2nd edn. New York: McGraw Hill, 1978, fig. 16.4, p. 210.

(A)

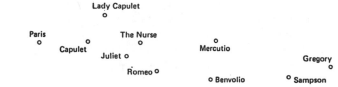

Tybalt

Lady Capulet

Paris   The Nurse

Capulet   Juliet   Mercutio   Gregory

Romeo   Benvolio   Sampson

Friar Lawrence

Friar John   Escalus   Balthazar

Montague   Abram

Lady Montague

(B)

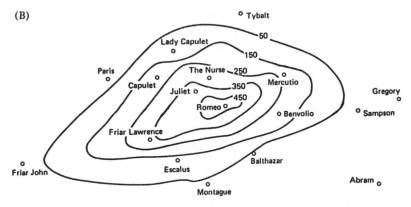

(C)

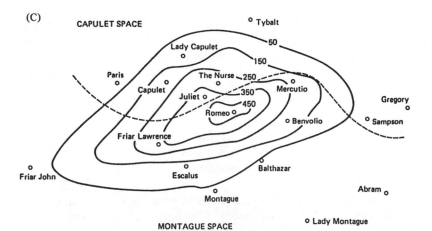

CAPULET SPACE

MONTAGUE SPACE

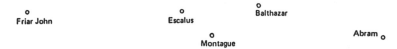

measuring attitudes between groups we can produce maps of hate or fear. The literature already includes one map of love; one Shakespeare play (*Romeo and Juliet*) has already been 'mapped' in terms of multi-dimensional scaling with Romeo and Juliet close together at the centre but the Montagues kept firmly separate from the Capulets (figure 3.5).[10]

SPACE AND KRONBERG

These example of non-linear mapping show that space is a more complex idea than a standard dictionary might suggest. Space means extent or area, usually expressed in terms of the earth's surface. It does not mean space in the sense of outer space (e.g. NASA, the National Aeronautical and Space Administration) nor space in the sense of arranging things in tidy rows. Space may be measured in terms of its length (as a separating distance) or its area (as territorial extent). It may be measured in familiar geometrical units (kilometres or miles), or in economic units (travel time in hours or travel costs in pounds) or in still other ways (such as psychological units in terms of familiarity or fear).

But in defining space it is also useful to define it as part of a trio of words – space, location, and place – which geographers use frequently and which, since they are also used in everyday language, can lead to a certain amount of confusion. Location means a particular position within space, usually a position on the earth's surface. Like the word space, it is rather abstract in meaning when compared to the third word in the trio. It can be measured in terms of a local Cartesian grid, or by reference to a spherical co-ordinate system of meridians and parallels.

Place also means a particular location on the earth's surface: but an identifiable location on which we load certain values. So a location becomes a place once it is identified with a certain content of information. Sometimes the content is a physical fact. For example, latitude 27° 59'N, longitude 86° 56'E is an abstract location which we only recognize as a place once we know it describes the position of Mount Everest, the highest point on the earth's land surface. In other cases, the information content is a human experience.

What gives a place its particular identity was a question which occurred to physicists Niels Bohr and Werner Heisenberg when they visited Kronberg Castle in Denmark. Bohr said to Heisenberg:

> Isn't it strange how this castle changes as soon as one imagines that Hamlet lived here? As scientists we believe that a castle consists only of stones, and admire the way the architect put them together. The

stones, the green roof with its patina, the wood carvings in the church, constitute the whole castle. None of this should be changed by the fact that Hamlet lived here, and yet it is changed completely. Suddenly the walls and the ramparts speak a quite different language. The court-yard becomes an entire world, a dark corner reminds us of the darkness in the human soul, we hear Hamlet's 'To be or not to be'. Yet all we really know about Hamlet is that his name appears in a thirteenth-century chronicle. No one can prove that he really lived, let alone that he lived here. But everyone knows the questions Shakespeare had him ask, the human depth he was made to reveal, and so he, too, had to be found in a place on earth, here in Kronberg. And once we know that, Kronberg becomes quite a different castle for us.[11]

Capturing this sense of place is one of the tasks of regional geography which I examine in the fourth essay in this book.

MEASURING GEOGRAPHIC SPACE

But before doing so, there is one more hurdle to jump. And since it is a higher one, readers already out of breath may well prefer to walk around it and go straight on to chapter 4.

Although geographers are concerned with a near-spherical planet, only a small proportion of their work is truly global. Cartographers and climato-logists alone used to grapple with surface of the sphere as an unbroken continuity: as the world economy integrates, economic geographers are starting to join them.[12] For the most part, geographers have concentrated their research on spatial models of small and confined segments of the earth's surface. Such segments may be thought of as a jigsaw of local areas, some sharp-edged and close-fitting, some with fuzzy limits and overlapping. These internal edges which split up an area of interest into sub-areas may be thought of as map infrastructure. Different jigsaws are used for different information; electoral districts for voting results, census districts for demog-raphic figures, watersheds for hydrological data and so on. Attempts to standardize this jungle of areas by adopting grid squares or using standard regions have met with limited success.[13]

Boundaries of districts can change from one time period to the next, and the problem arises of how to patch together a consistent time-series of observations for each unit. This plagued my own research with Andrew Cliff on tracking the geography of the spread of epidemics in Iceland.[14] Iceland has the most complete epidemiological record both in time and

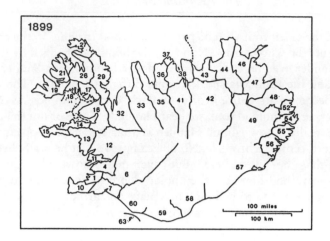

1899

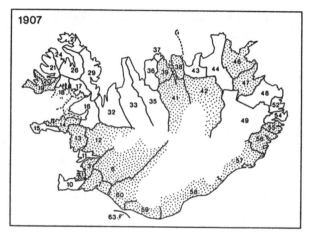

1907

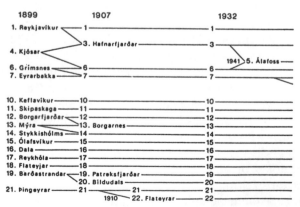

1899	1907	1932
1. Reykjavíkur	1	1
	3. Hafnarfjarðar	3
4. Kjósar		
		1941 5. Álafoss
6. Grímsnes	6	6
7. Eyrarbakka	7	7
10. Keflavíkur	10	10
11. Skipaskaga	11	11
12. Borgarfjarðar	12	12
13. Mýra	13. Borgarnes	13
14. Stykkishólms	14	14
15. Ólafsvíkur	15	15
16. Dala	16	16
17. Reykhóla	17	17
18. Flateyjar	18	18
19. Barðastrandar	19. Patreksfjarðar	19
	20. Bíldudals	20
21. Þingeyrar	21	21
	1910 22. Flateyrar	22

space of any country in the world. Data are available for more than thirty common infectious diseases, as well as for other non-communicable diseases on a monthly basis for some fifty geographical areas back to 1896, and more irregularly for over another century before that. The basic reporting unit for these data is the medical district (*leaknishera*). But over the last ninety years there have been many changes both in district numbers and in boundary locations. In order to bring these districts to a common geographical mesh, we worked with the smallest parts of the Icelandic jigsaw, the *hreppr*. (This is roughly equivalent to an English parish and to an American township.)

The medical districts we were studying are composed of *hreppr* and legal documents exist which define which *hreppr* make up which medical districts. Using these legal documents and a base map it was possible to prepare a complex branching diagram of areas, part of which is shown in figure 3.6. This shows all the known changes in the medical district boundaries and their date of occurrence. Thus we can see that medical district 1, Reykjavikur, existed throughout the study period but that district 3, Hafnarfjardar, was first created in 1907 by an amalgamation of Kjosar (district 4) with part of Reykjavikur. The branching diagram may be converted into a geographical format to produce maps of the boundaries of the medical districts.

Thus spatial continuity and temporal continuity represent irreconcilable goals in geography. If we are to preserve a consistent time-series, then we need to sacrifice (through amalgamation) a great deal of spatial detail. Conversely, if we wish to retain the maximum amount of spatial detail then we can only have very short and broken time-series.

Even when shifts over time are not an issue, the impact of internal boundaries on quantitative measures may be severe. Figure 3.7 illustrates some findings by Stanley Openshaw and Peter Taylor[15] on age and voting in the United States. In a study of the association between the Republican vote and older populations for the state of Iowa, USA, they obtained correlation coefficients ranging from a high of +0.86 to a low of +0.26 when the hundred counties that make up the state were grouped, using different criteria, into six conventional regions. When many other six-region partitions were tested, the range of possible values for the correlation coefficient widened still further to include negative as well as positive associations.

Figure 3.6   The instability of geographical data infrastructure. Changes in Icelandic medical districts from 1899 to 1907 shown both as maps and as a branching diagram.
Source: A. D. Cliff, P. Haggett, J. K. Ord and G. R. Versey, *Spatial Diffusion: An Historical Geography of Epidemics in an Island Community*. Cambridge: Cambridge University Press, 1981, fig. 4.2, pp. 60–1.

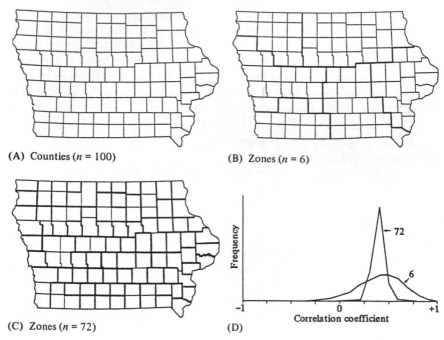

(A) Counties (*n* = 100)   (B) Zones (*n* = 6)

(C) Zones (*n* = 72)   (D)

Figure 3.7 Impact of regional boundaries on statistical results. Openshaw and Taylor's findings on the frequencies of correlation coefficients at different scales for two variables within Iowa. The maps show a single example of the very large number of boundary configurations at each spatial scale. The maps show the original counties (A) and one example of a six-region grouping (B) and a 72-region grouping (C). Many different spatial arrangements of these two types of grouping are possible, and they lead to different correlation coefficients, as shown in diagram (D).

Source: Peter Haggett, 'The edges of space'. In R. J. Bennett (ed.), *European Progress in Spatial Analysis*. London: Pion, 1981, fig. 3.5, p. 60. The original work was reported in S. Openshaw and P. J. Taylor, 'A million or so correlation coefficients: three experiments on the modifiable unit area problem'. In N. Wrigley (ed.), *Statistical Applications in the Spatial Sciences*. London: Pion, 1972, pp. 127–44.

The general lesson is that geographical analysis cannot be pulled clear of the space within which it operates. Canadian geographer Leslie Curry has argued that the attenuating effect of distance, as measured by spatial interaction models of journey-to-work or migration, confounds two distinct components: a 'behavioural' component describing how trips are determined by the cost of distance, and a 'map pattern' component describing the particular jigsaw between which trips are occurring.[16] In most models we measure a single term which mixes up both effects. As with the Iowa voting study, results are influenced by the spatial arrangement within which we study a problem.

LOOKING OVER THE EDGE

If geographers are working within the cosy confines of one well-defined segment of the earth's surface, then their problems too are well defined. But what happens if we wish to extrapolate, to peer over the edge of our region into the world outside? To what extent can we use our local knowledge and generalize it out to surrounding areas?

Although the problem is mathematically complex, we can illustrate it with a rather simple illustration. Let us begin with a contour map for all the observations inside our study region as in figure 3.8A.[17] Geologist Bill Krumbein has examined the problems to be met in extrapolating a trend-surface map beyond the control limits (areas where data are known) into the peripheral boundary zone (areas where data are unknown).[18] To do this, a rectangular section from the McClure, Pennsylvania, topographic map was sampled to give 90 data points recording their elevation and location. As shown in figure 3.8A, this control area is traversed diagonally from SW to NE by a long ridge which extends laterally beyond the mapped area.

Two types of trend-surface map were fitted to the elevation data. First was the commonly used 'polynomial' model which structures the data in such a way that a succession of fitted surfaces (linear, quadratic, cubic etc.) can be constructed. Second was the less common 'Fourier' model which structures the data to produce a series of harmonic surfaces, with wave forms of diminishing length as the order of the surface rises. For our purposes the maths can be left to one side, for the maps they produce tell the story.

Maps B and C show the application of the two models to the test area. In each case the shape of the surface is estimated for the control points lying within the rectangular control area but the surface extends into the surrounding boundary zone. The degree of fit achieved within the control area is given for each model. For both models, the fit increases as the order of the surface (and the complexity of the generating equation) increases. Thus in the polynomial model, the quartic surface with 15 coefficients in the equation gives better fits than the cubic surface with 10, or the quadratic surface with 6. The more you pay for your carpet, the better the quality.

Most interest focuses on the performance of the maps in the boundary zone. From looking at figure 3.8, two things stand out. First, the low-order quadratic map gives a more satisfactory representation of the continuing

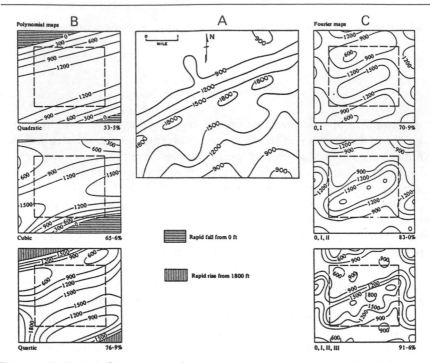

Figure 3.8　Extrapolating across the map edge. W. C. Krumbein's investigation of the performance of trend-surface models in boundary areas. Contrasts between the estimates of polynomial models (left) and Fourier models (right) in extrapolating contours beyond the edges of the control area. The original contour map (the control area) is shown at an enlarged scale in the centre.

Source: Peter Haggett, 'The edges of space'. In R. J. Bennett (ed.), *European Progress in Spatial Analysis*. London: Pion, 1981, p. 57, fig. 3.3. The original work was reported in W. C. Krumbein, 'A comparison of polynomial and Fourier models in map analysis'. Office of Naval Research, Geography Branch, ONR Task No. 388-078, Contract 1228(36), TR-2. Evanston, Illinois: Northwestern University, 1966.

diagonal ridge than its higher order equivalents; in the latter the extrapolated values rapidly increase or decrease beyond reasonable numerical values (shaded areas). Paradoxically, the better the fit of a polynomial surface within the control area, the poorer it is as an extrapolatory model outside the map boundary. Second, the Fourier maps differ markedly from the polynomial in that they simply display periodic repetition of the pattern within the control area when extrapolated. This generalization applies to Fourier maps of all orders, so that extrapolation of a Fourier surface, even one of low order, seems ruled out beyond the edge of the control map.

IRREGULAR CONTROL POINTS

One of the critical spatial factors in determining the accuracy of any contour maps is the geographical location of our observations. For example, David Unwin and Neil Wrigley have re-examined the role of the spatial distribution of control points in determining the form of these polynomial surfaces

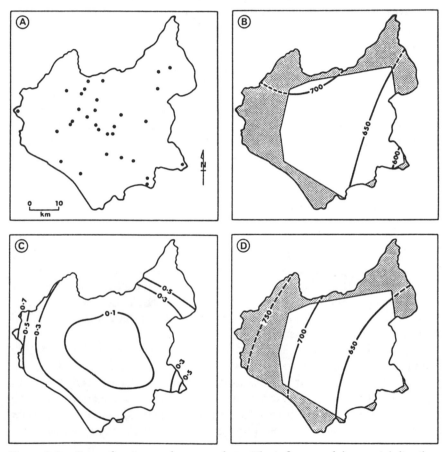

Figure 3.9   Control points and map surfaces. The influence of the spatial distribution of critical control points on trend surfaces fitted to precipitation values over an English county, Leicestershire. (A) Rainfall stations. (B) Trend-surface map based on the station data. (C) Relative influence of each station on the trend-surface map; the higher the value, the greater the influence. (D) Change in the trend surface map when data from one critical station (that located furthest west in map (A)) is ignored.

Source: D. J. Unwin and N. Wrigley, *Institute of British Geographers, Transactions*, New Series, Vol. 12 (1987), pp. 147–60; figs 10 and 11, pp. 158–9.

as used by Krumbein.[19] They illustrate this by a study of spatial trends in rainfall across an English county, Leicestershire (see figure 3.9). Mean annual precipitation totals are available from the rain gauge locations shown in map A. A fitted linear surface gives a weak but highly significant 'shedroof' spatial trend showing rainfall declining from northwest to southeast. When the next highest surface (the quadratic) is fitted, the fit improves from 34 to 44 per cent. This is shown in map B. If, however, the most westerly rain gauge in the county (at a location on the boundary) is omitted and the quadratic surface is recomputed, then the resulting surface has a different shape (D).

Clearly, the omitted observation is exercising 'leverage' on the surface, and Unwin and Wrigley go on to provide a method for measuring this effect. Map C shows the leverage values obtained for the different Leicestershire locations with high values located around the margins of the map. Parallel edge effects arise when geographers try to study all sorts of spatial distributions and special techniques have to be developed to cope with them.[20]

DECODING SPATIAL MESSAGES

So far in the second half of the chapter we have been concerned with building maps. But map making is only one half of what is sometimes termed the 'mapping cycle'. This envisages a circular process in which field data are converted into maps (map making) which are then interpreted by others to give new information or insights about the field (map interpretation). The first part is sometimes described as the 'encoding' of spatial information, the second as 'decoding'.

We can illustrate what is meant by decoding by considering a specific example.[21] Consider figure 3.10, which shows the incidence of an unpleasant blood disease (toxoplasmosis) in cities in the Central American country of El Salvador. Measured rates are expressed as deviations from the national incidence (the average of the rates for all cities) in map A, so that a city showing -0.04 has a rate four per cent lower than the country as a whole.

Figure 3.10   Error and confidence in statistical mapping. Use of Bayesian estimators by Efron and Morris in mapping of (A) the distribution of toxoplasmosis in thirty-six locations in El Salvador. Map (B) shows contour maps of incidence based on local mean values and map (C) modified contour maps based on James-Stein estimators in which the variability at each location has also been taken into account.
Source: Peter Haggett, 'The edge of space'. In R. J. Bennett (ed.), *European Progress in Spatial Analysis*. London: Pion, 1981, fig. 3.6, p. 62.

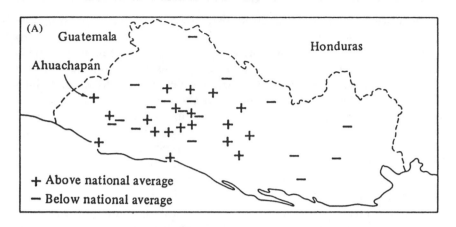

(A)
Guatemala
Ahuachapán
Honduras

+ Above national average
− Below national average

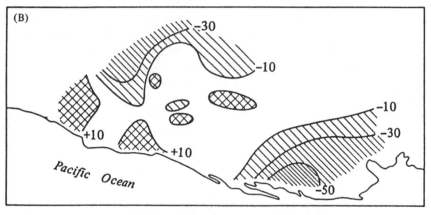

(B)
−30
−10
−10
−30
+10
+10
−50
Pacific Ocean

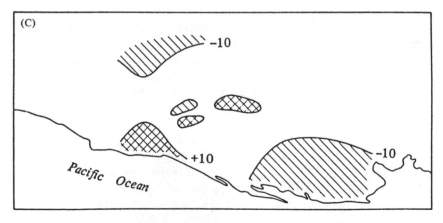

(C)
−10
+10
−10
Pacific Ocean

One obvious map to draw from these data is shown in map B. The incidence value for each city is taken as the best estimator and used to generate a contour map of the disease for the country. But how accurate is that map? If we go back to the original data we find that tests were carried out for toxoplasmosis on only five thousand people in El Salvador (which has a population of 4.8 million). Nor was the sample drawn equally from the 36 cities. Further, the cities showed extreme contrasts in the number of patients examined and variability of the disease; cities with the very high and the very low incidence were those with largest variability. Given these facts, map B takes on a less confident shape. We can no longer be sure that the highs and lows are a result of real spatial variations in the disease or just a by-product of the sampling process.

Statisticians have provided methods by which these complications can be taken into account. For example, map C shows a revised map based on such a measure called the James-Stein estimator. In some cases the change in map values is considerable; thus the western city of Ahuachapan, ranked highest with a value of 29 per cent above the national value, is demoted by the James-Stein estimator to twelfth ranking position, only 5 per cent above the national value. The establishment of more accurate map patterns from spatially distributed data has a long history in quantitative geography going back to Choynowski's work on cancer in Poland.[22] Recent work on statistical estimation theory promises to provide a whole new family of estimators (of which the James-Stein statistic is simply an example) on which more accurate maps can be drawn.[23]

SPACE AS RESOURCE AND LIABILITY

We have seen in this essay a few of the ways in which geographical space enters into a geographer's calculations. In some, it acts as a marvellous resource allowing new insights from information that might otherwise be regarded as unstructured and unintelligible. The possibilities for non-linear mapping that the high-speed computer opens up have greatly aided this process of unravelling complex space. In others, space acts as a liability, reducing our freedom for analysis and bringing all sorts of side effects into what are otherwise straightforward issues. Where space acts as a liability there appear to be three continuing tasks for geographers. First is the precise identification of spatial effects separating those that are significant from the trivial. Second is the careful development of 'space-proofed' analytical methods which build in corrections. Third, we must flag those areas where errors may be large and solutions may not be feasible.

Recent decades have shown encouraging signs of renewed interest in these problems both from inside and outside geography.[24] Balancing the space-correcting tasks within quantitative geography are those which exploit spatial effects as one of the central planks of geographical research. These call for the identification and exploitation of those areas where the spatial aspect, one area of especial concern for geographers, contributes insights both to the encoding and the decoding of map information.

New alliances with colleagues in mathematics, and in particular with those in statistics, have needed to be forged and this chapter has had to step lightly over the foothills of some high and difficult mathematical ranges. Mathematics and Geography have been harnessed together for more than 2,000 years. Even the oldest national geographical society in America – the American Geographical Society – was originally founded as the American Geographical and Statistical Society of New York. These recent alliances mark a return to old roots.

4

Regional Synthesis

'When you say "hill,"' the Queen interrupted, 'I could show you hills, in comparison with which you'd call that a valley.'
Lewis Carroll, *Through the Looking Glass and What Alice Found There* (1872)

In geography, the individual, the idea and the region are rarely far removed. Thus we think of the Swedish geographer, Torsten Hägerstrand, and his idea of innovation waves springing from his research in Central Sweden. Or Donald Meinig and land-use cycles in Southern Australia. Or Walter Christaller and central-place hierarchies in Bavaria. Or Sidney Wooldridge and erosion cycles in the English Weald. The full list would encompass much of the best geographical writing.[1]

For me, one of the most compelling of these associations comes in the book written by a young German economist at Kiel University. August Lösch had died at the end of World War II at the age of only thirty-nine, his death partly the result of a strength of character which forbade any compromises with the National Socialist regime. His book on the *Economics of Location* was a classic, but difficult, examination of the relation between economic goods and geography;[2] not only was the text interesting, but Lösch peppered it with footnotes raising intriguing ideas which he couldn't solve. I had gone up to Cambridge much more interested in physical geography than in its social science or regional counterparts, but Lösch's book diverted me in both senses of that word.

I found it strikingly original in that he asked significant new questions about the role of space in the economy. Thus he saw the spatial pattern of cities, towns and villages almost in the manner of a solid-state physicist (figure 4.1); each settlement like an atom bonded together in a matrix of molecular interconnections. But Lösch was also blessed with a sense of tradition and history which allowed him to blend in past locational contributions from von Thünen onwards and to place his results in a solid regional framework. The book draws heavily on Lösch's studies in the United States in the 1930s, particularly from the Mid West. But while he acknowledges his American debt, he reflects:

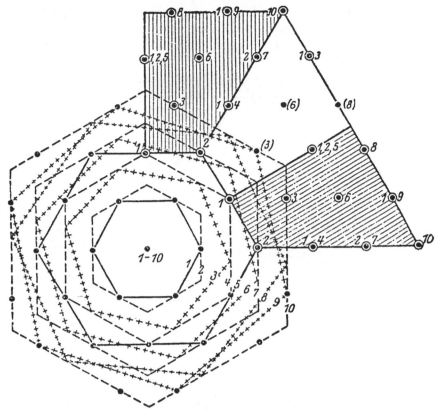

Figure 4.1   Lösch's 'solid state' model of the tributary areas of settlements. Ten smallest economic areas with section with many towns hatched.

Source: A. Lösch, *The Economics of Location.* New Haven: Yale University Press, 1954, fig. 27, p. 118.

My youthful experiences in a little Swabian town constitute the real background of this book. I am convinced that we rarely learn to know any conditions as intimately as those among which one grew up. We can judge with certainty only a small understanding and familiar world like this, and we transfer the findings to large problems afterwards. My Swabian homeland constitutes such a world in miniature.[3]

Lösch's comment struck a familiar chord. I was exceptionally fortunate in that my father introduced me to the landscape he loved around our Somerset home. Both family finances and wartime petrol rationing ensured that we followed Carl Sauer's dictum that 'locomotion should be slow; the

slower the better'.[4] This small piece of countryside, like Lösch's native
Swabia, seemed to illustrate many of the major locational questions and
even to hint at their answers. Later, larger and more rapidly traversed

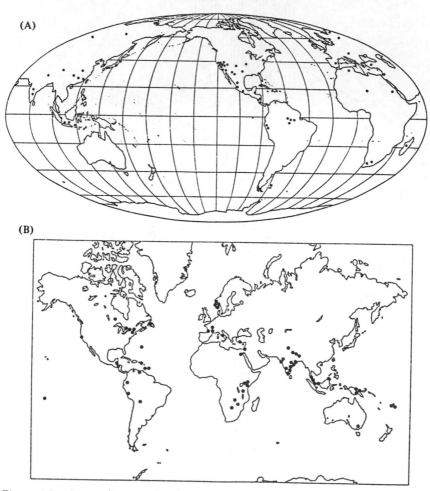

(A)

(B)

Figure 4.2   Areas of geographical research for two leading geography departments.
(A) Field areas of postgraduate students from the Department of Geography,
University of California at Berkeley. (B) Field areas outside Great Britain of
undergraduates at the Department of Geography, University of Cambridge, 1986–
88. Note that, since the Cambridge map cuts out local fieldwork, the two maps are
not directly comparable.

Sources: Map (A) is based on map in *Berkeley Georgraphy 1988*. Berkeley: University of
California, Department of Geography, 1988, p. 5. Map (B) is based on *Geography at
Cambridge*. Cambridge: University of Cambridge, Department of Geography, 1988, p. 12.

landscapes in southern Europe, in Brazil, in the western United Sates and – more recently – in South East Asia and the Pacific have been grafted on to this slow-grown native rootstock.

SELECTION OF REGIONS

Just how any one geographer ends up studying one particular region is something of a mystery. Where we study at university and the constraints of language and accessibility must play some part. Indeed, systematic studies of the location of field areas tend to confirm both a distance decay effect and a language-cultural effect; the first leading to a heavy concentration of research on local, accessible regions; the second to the choice of overseas areas with which there is a common language. As figure 4.2 suggests, both are different aspects of a least-effort solution.

My own experience suggests that both random and systematic elements influence the selection of a research area. Undergraduate work in Portugal gave me a smattering of that language so that when, at University College, London, Clifford Darby asked me to give some South American lectures it was not the Spanish-speaking part but Portuguese-speaking Brazil which attracted me to fieldwork. Some of that fieldwork has been described in the second chapter.

Serendipity played an even greater part in the choice of the county of Somerset for a recent study of rural service changes; indeed, attending a colleague's funeral must be among the more unlikely reasons for regional choice.[5] The exigencies of wartime service played some part in the regional specialisms of at least two Cambridge geographers: Ben Farmer's work on Sri Lanka dated from army service in that island, while Charles Fisher's knowledge of south-east Asia was sharpened by internment in the Japanese prisoner-of-war camp at Changi.[6]

But geographers do not rely wholly on happenstance. For example, the choice of Iceland for my work over the last decade with Andrew Cliff[7] was the result of a rational selection procedure (see figure 4.3) in which more than a dozen potential study areas were whittled down to one.

As we wrote:

The fact that we found ourselves studying . . . a single island community (Iceland) over an eighty year period was not a matter of accident or intuition. Rather it was forced on us by the contribution we wanted to make to spatial diffusion studies. Iceland is, despite its intrinsic interest, essentially a means to this end.[8]

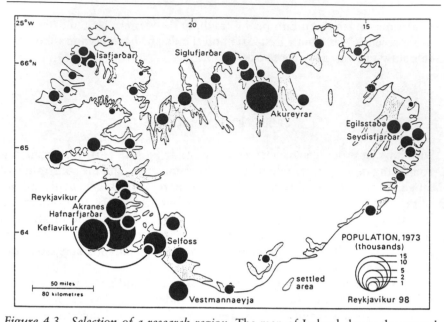

Figure 4.3   Selection of a research region. The map of Iceland shows the coastal
distribution of the main population centres with settled areas stippled.
Source: A. D. Cliff, P. Haggett, J. K. Ord and G. R. Versey, *Spatial Diffusion: An Historical
Geography of Epidemics in an Island Community.* Cambridge: Cambridge University Press,
1981, fig. 13.2, p. 52.

Our regional selection process argued that, for any empirical study of
spatial diffusion to yield useful tests of spread models, it must meet four
conditions – replicability, stability, observability and isolatability. Replicability meant repeating diffusion waves and not a single event. Stability
implied that the phenomena being spread in these repeating waves must
remain unchanging over time. Observability meant that our models needed
good, reliable data with an intricate matrix of space and time observations.
Finally, we needed our field area to be so sufficiently isolated that it was not
contaminated by random waves coming in from other areas.

One by one we had to reject most of the topics which had been studied
previously in geographical research on diffusion as they failed our four tests.
Thus social innovations were found to be non-replicable, economic fluctuations unstable over time, and so on. At the end of the search we found that
only the waves set up by recurrent waves of human infectious diseases were
likely to be suitable for the modelling we had in mind. Even here, further
narrowing down had to occur and we found ourselves studying a very
restricted and specialized field – Type II measles waves. (For the form of
these waves turn on to figure 5.4 in the following chapter.)

In practice, Iceland produced a number of remarkable bonuses. It was not only isolated but, as figure 4.3 shows, the individual settlements were arranged like a string of beads. They formed an encircling archipelago of limpet-like settlements around the littoral of the island and, in terms of disease contact, were each rather separate from each other (plate 7). The detail and range of the country's public health and demographic records were exceptionally high, even by high Scandinavian standards (and well above that for North American or Western European countries). We were well content with our choice.

EXPERIMENTS OR ACCIDENTS?

But even if we choose our areas carefully, formal experimental designs in the sense used by statistical theorists are rarely possible in geography.[9] Occasionally experiments may be approximated by making field observations in a controlled test.

In the Brazilian work described in an earlier essay it was possible to identify areas for sampling by the use of factor-combination regions. The principle of these regions is shown in figure 4.4A. Areas where three factors are present or absent are shown schematically, together with the appropriate two-factor and three-factor combinations. The lower map, B, shows the application of this principle to the Fortaleza basin in south-east Brazil (compare with figure 2.2 in the earlier essay). Here the four factors referred to different environmental factors — terrain (a), soils (b), farm size (c) and farm accessibility (d) — which were thought to have a major influence on land use.

Formal selection of field areas has a place in regional studies but grasping opportunities where and when they arise is equally critical. Again Iceland provides one such example where an historical event gave the opportunity for a rare epidemiological experiment.

In looking at the Iceland measles waves which swept the island from the turn of the century, Andrew Cliff and I turned first to the 1904 wave which affected just the northwestern part of the island.[10] The story of its spread and containment in the small farms and fishing villages of that barren fjord country is not without interest, but from the experimental viewpoint the critical events took place in two neighbouring parishes (called Eyri and Ogur) on the southern side of the main fjord (see figure 4.5). The little Lutheran parish church at Eyri (see plate 9) was crowded on Saturday 21 May 1904 for the annual confirmation ceremony for those children of the parish who were around fourteen years old. The following day, Whit Sunday, the travelling Lutheran minister went on by boat to the

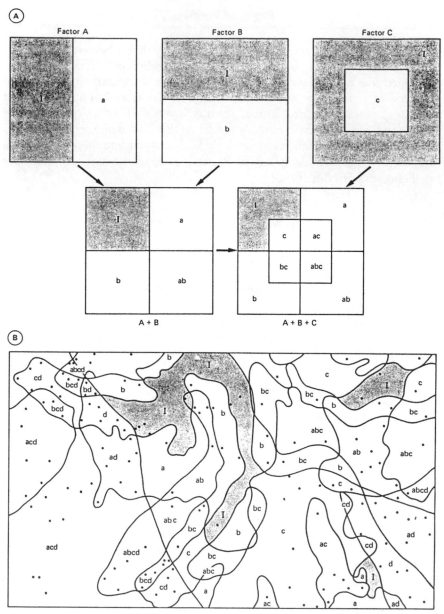

Figure 4.4 Factor combination regions. (A) Subdivision of a hypothetical region using combination of three factors, a, b, and c. Areas where factors are absent are denoted by I. Factors b and c are successively imposed on a. (B) Location of sampling points within factor combination regions in the Fortaleza Basin, Taubaté county, Brazil. Compare with plate 4.

Source: Peter Haggett, Andrew Cliff and Allan Frey, *Locational Analysis in Human Geography*, 2nd edn. London: Edward Arnold, 1977, fig. 8.6, p. 275.

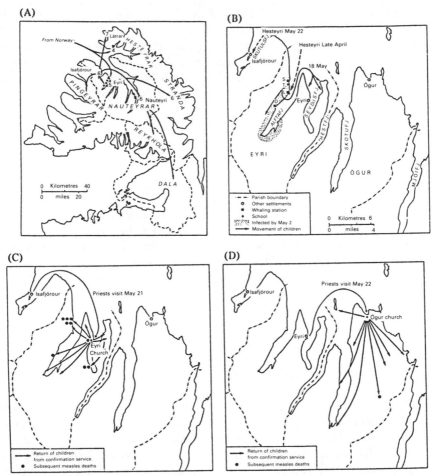

Figure 4.5   Test and control regions. (A) Spread of measles in the summer 1904 epidemic. (B) The two fjord-side parishes of Eyri (the test area, left) and Ogur (the control area, right). (C) and (D) show the higher rate of deaths in the test parish as compared to the control parish.

Source: Andrew D. Cliff, Peter Haggett and Rosemary Graham, 'Reconstruction of diffusion processes at different geographical scales: the 1904 measles epidemic in northwest Iceland'. *Journal of Historical Geography*, Vol. 9 (1983), pp. 347–68; fig. 4, p. 35.

neighbouring church at Ogur and carried out a similar ceremony. Both parishes had similar populations and age structures at that time; in both the small farms were scattered on occasional patches of flat land along the sides and heads of the fjords. There was little to separate the geographical character of the two parishes.

But from the experimental point of view one difference was crucial. At least one of the children at the Eyri (Saturday) ceremony was infectious with measles, while at Ogur (Sunday) this appeared not to be so. The maps in figure 4.5 show what then happened. The subsequent explosion of cases in the one parish (with consequent deaths) and the relatively slow spread in the other provided a comparative test that arose from accident rather than design.

THE ILLAWARRA INCIDENT

A similar opportunity for geographic study arising from an Act of God came on Saturday 5 January 1975, when at 9.30 p.m. the Australian city of Hobart was cut in two. The capital city of Tasmania, Hobart had a population at the time of 145,000, spread along the eastern and western sides of the Derwent River (see plate 10). On the western side lay the old city of Hobart with 104,000 people; on the eastern side the fast-growing suburb of Clarence with 40,000 people and the international airport. Across the kilometre-wide river was the four-lane highway over the Tasman bridge joining the city to eastern suburbs and the airport.

The disaster which struck the bridge was so sudden that a few cars ran on into deep water. The *Lake Illawarra*, a freighter loaded with zinc concentrate, collided with the bridge pillars. The collision brought down a major section of the bridge decking which fell on to the freighter, sinking it under the impact.

On a normal weekday 34,000 people crossed the bridge in each direction and with its closure the only alternatives for commuters were to go upriver to the next bridge on a thirty-mile diversion, or queue for the few old and overcrowded ferries that were pressed back into service. This unwelcome but opportune experiment on the effect of transport dislocation on the economic and social geography of a city has been monitored by the department of Geography at the University of Tasmania. They were able to show the impact on economic life (from changes in retail sales to house prices) and social life (from rising divorce rates to depleted club membership) of this event and how adjustments were made.[11] It presented a unique experimental opportunity for the regional study of a divided city in the same way that sudden natural disasters (such as floods and hurricane surges) allow opportunity for insights into environmental adjustment.

RATIONALES FOR REGIONAL STUDY

So far we have used 'region' in a loose way to indicate a focus on one part of the earth's surface. We now need to tighten that up. The central role of the regions has been so widely accepted within the geographical discipline that asking a geographer why he studies regions is like asking a Christian why he studies the Gospels. A halting response suggests that often the most fundamental beliefs are the hardest to define. But if we stand back and examine the reasons why we do what we do, then several rationales for regional study, rather than a single one, emerge. I have identified five reasons – regions as exemplars, as anomalies, as analogues, as modulators and as covering sets – but this is not an exhaustive list.[12] Let us take each in turn.

First, regions serve as exemplars. They give local substance to generalization, put flesh on the logical structure, provide a specific example to press home an argument. Thus, *The Climate of London* provides an illustration of the modifying effects of a large urban area on the atmospheric envelope covering it, translating the general physics of boundary-layer climates into the local experience of smogs and gusty streets.[13] The exemplar may be used not only to illustrate principles, but also to illuminate a wider area. A classic example of using micro-regions to illustrate the characteristics of a much larger geographical area is provided in Platt's *Field Approach to Regions*,[14] where a single coffee fazenda throws light on some aspects of the whole Brazilian coffee belt.

Second, and in contrast to the first argument, regions serve as anomalies or residuals. In this case the purpose is to underscore how a local part of the earth's surface departs from a general statement or relationship. Thus the distinctiveness of the 'driftless' region of south-west Wisconsin lies primarily in its anomalous character in relation to the glaciated areas on all sides. Its most striking feature is a negative one, the *absence* of glacial deposits, and this only makes sense in respect of broader generalizations about an area in which glacial deposits dominate half a continent.

Anomalies and residuals play an important role in testing and reformulating general models. The failure of a general explanation to make sense in a specific region may be due to limitations in the model itself, or the fact that several influences are coming together in such a way that an expected effect is either heightened or reduced. Trewartha's *Earth's Problem Climates* takes advantage of this approach in picking out for special consideration those areas which have observed characteristics not expected on the basis of general models of the earth's atmosphere.[15] We might expect, on most general circulation models, that the Amazon Basin would have high rainfall. Why then, asks Trewartha, are parts of it rather dry?

THE BRIDGE ON THE RIVER KWAI

Third, regions may be studied as analogues. This describes the matching of

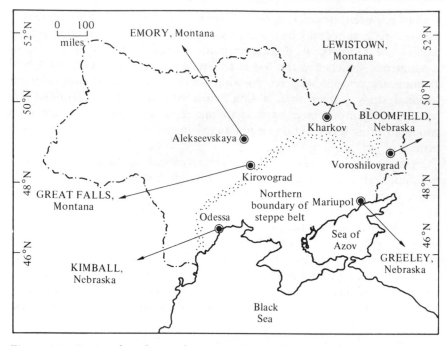

Figure 4.6   Regional analogues for crop exchange. Location of climatic stations in the Ukrainian SSR. For six sample stations, the closest climatic analogue in the United States is shown. The technique was used in experiments with Ukrainian wheat varieties to find 'matching' climatic areas within the Great Plains wheat belt.
Source: Map based on tables in Michael Yakovlevich Nuttonson, *Ecological crop geography of the Ukraine and Ukrainin agro-climatic analogues in North America*. Washington, DC: American Institute of Crop Ecology, 1947.

the characteristics of one region to matching regions in one or more other parts of the world. We saw a teaching example of this in figure 1.2 and a simple research example is provided by a primitive study I once made of the cork oak (*Quercus suber*).[16] This is a type of oak tree native to the western Mediterranean basin. By identifying the trees and the climatic environment in its home range, it was possible to pinpoint, on climatic grounds, areas of western North America where it might also flourish. This illustrates the distinction between the 'benchmark' region from which the characteristics are measured, and the 'analogue' region constructed from the spatial transfer of the characteristics. Many illustrations of the method are given in the work of Nuttonson's group at the Institute for Crop Ecology in Washington, DC.[17] Typical is their report on the climate of the Ukraine region of Soviet Russia specified in terms of the critical parameters for growth of grain crops. Climatic stations in the United States are then searched to identify those most closely matching the Ukraine. Figure 4.6 shows the resulting regional comparisons between the two areas.

A similar search for regional analogues has been followed by film producers. Thus Sam Spiegel's *Bridge on the River Kwai* was ostensibly set in Burma, but actually shot on location in the dry zone of south-east Sri Lanka (plate 10). Carlo Ponti's *Doctor Zhivago* was not filmed on the Russian steppes but in various 'matching' parts of the world – Spain,

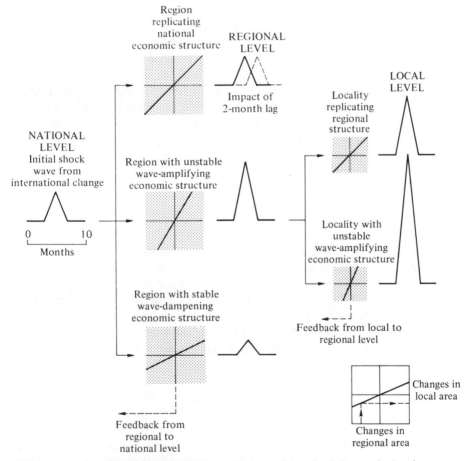

Figure 4.7 Hierarchic transmission of shocks through regions. Idealized representation of shocks from a national economy being transferred to individual regions with different response patterns.

Source: P. Haggett, *Geography: A Modern Synthesis*, 3rd revised edn. New York: Harper and Row, 1983, fig. 24.7, p. 593.

Ontario and Finland. One extreme case of a low-budget remake of *Beau Geste* substituted the cold dune ridges of Scotland's Culbin Bay for the shimmering sands of the Sahara.[18] Perhaps I should add a further litmus test for recognizing a latent geographer – the member of the cinema audience who stays on in the emptying theatre to see if the last line of the credits roll will answer 'Where in the world was it *really* filmed?'

REGIONS AS MODULATORS

Fourth, regions act as modulators: their unique structures modify the ways in which a region behaves over time. Studies of the geographical pattern of business activity show that local regions have activity cycles – in employment, investment, housing starts, inflation and the like. In some degree the local economy follows the trends in the larger national economy within which it is set. But they do so in a regionally modified way. Those which have a large concentration of industries which are cyclically vulnerable, such as steel and automobiles, may show a marked over-reaction to national trends. Other areas which have rather stable employment, such as government administration or higher education, may show a muted response to the same trends. As figure 4.7 shows, we can conceive the local economy of a given region as modulating a cyclic pattern.

While geographical applications have been relatively few so far, the concept of the region as a modulator could in theory be extended. The prerequisite conditions are that a region shows distinctive behaviour over time and that its behaviour is logically connected in two ways: first, vertically, to oscillations of the larger section of which it is a part; and second, horizontally, to other regions. The principles used in regional economic models could be extended to show how the sediment yields of a small river basin relate to the overall river basin of which it forms a part, or to the political reaction of different electoral regions to national swings in political opinion.[19]

REGIONS AS JIGSAWS

A fifth use of regions is as 'covering sets': this is the term used to describe the jigsaw of areas which completely exhaust a continent or country. The need to reduce the complexity of understanding of an overall system leads geographers to divide their spatial fields into smaller and more readily comprehensible sections. These may be world regions at one scale, major parts of continents or tiny divisions of a small valley at another. The problem is directly comparable with the historian's use of discrete historical periods, the playwright's use of acts and scenes, and the librarian s use of classification systems.

In each case, we need a set of rules to establish the protocol for placing an area into a given region, just as a librarian classifies each incoming volume. The rules need to be comprehensive enough to ensure that the whole global region is covered, i.e. that there are no books which we cannot classify. As we would expect, there will be considerable debate over the rules and much anguish over those difficult cases which lie on a boundary and are hard to classify.

SCALE IN REGIONAL SYSTEMS

The fact that the scale resolution problems we met in chapter 2 have long troubled geographers is plainly shown in the series of attempts that have been made to define regions in terms of a size hierarchy. Table 4.1 shows a series of terms used in the description of regional systems.[20] In the case of physiographic regions, the system applied by Fenneman in 1916 to the landform divisions of the United States (with its recognition of 'major divisions', 'provinces', 'sections and districts') was taken up by other writers. Unstead's 'systems of regions' filled in at the smaller levels the system Fenneman began at the larger. Linton integrated both preceding

Table 4.1 Comparative scales and terminology of regional systems

Approx. size (sq. mls)	Fennemann (1916)	Unstead (1933)	Linton (1949)	Whittlesey (1954)	Map scales for study
10^{-1}			Site		
10		Stow	Stow	Locality	1/10,000
10^2	District	Tract	Tract		1/50,000
				District	
10^3	Section		Section		
		Sub-Regn			
				Province	
10^4	Province		Province		1/1,000,000
		Minor Regn			
10^5	Major Divn		Major Divn	Realm	1/5,000,000
10^6		Major Regn	Continent		

Source: R. J. Chorley and P. Haggett (eds), *Frontiers in Geographical Teaching*. London: Methuen, 1965, table 5, p. 171.

systems in a seven-stage system which ran through the whole range from the smallest unit, the 'site', to the largest, the 'continent'.[21] Whittlesey presented a 'hierarchy for compages' with details of the appropriate map scales for study and presentation and followed this with a model study on Southern Rhodesia to illustrate his method.[22] But the decades since he made the call to 'fill this lacuna in geographic thinking' have not seen any rush to adopt the Whittlesey scheme.[23] Indeed I am only aware of one paper, that by the Southampton geographer, James Bird, which subjected Whittlesey's scheme to field testing.[24] Bird's two-scale comparison of the western peninsulas of Brittany and Cornwall suggested that, while a general (or small-scale) study showed the two areas to be similar, the intensive (or large-scale) study showed that the two peninsulas were quite dissimilar in most details. I find it surprising that Bird's deft illustration of a fundamental and very common geographical problem should have attracted so little comment.

The second major move in the period since Whittlesey's papers came from Canadian Allan Philbrick who published a very full scheme based on the concept of a sevenfold hierarchy of functions.[25] Corresponding to each function is a nodal point with its functional region. Here scale is introduced through the 'nesting' concept with each order of the hierarchy fitting within the next highest order. As a theoretical model Philbrick illustrates the case where each central place of a given order is defined to include four central places of the next lower order. This gives a succession for a seventh-order region of four sixth-order places, sixteen fifth-order places, and so on down to the final level of 4,096 first-order places. Although his attempt to apply this scheme to the eastern United States with New York and Chicago in the role of seventh- and sixth-order centres was only partly successful, the introduction of a scale component into a system of nodal regions has important theoretical implications. As in our view of the heavens, we see more and more stars – more and more regions – as the resolving power of our instruments improves.

WRITING REGIONAL GEOGRAPHY

The most difficult type of regional geography to write is that shown by the focus of rays in Hartshorne's diagram in chapter 1 (see figure 1.4). This involves the balance, integration and synthesis of information of different kinds from the systematic fields. Writing as a practitioner, John Paterson is able to identify what he sees as the six leading problems facing the regional geographer: the logical impossibility of providing a complete regional description in verbal form; the problem of identifying the regions them-selves; the problem of handling scale variations in presentation; the growing shortage of subordinate sub-regional materials; the submergence of regional distinctiveness; and the limited amount of innovation possible.[26]

All the first three factors are timeless and may only have become marginally more difficult over the last few decades. This fourth factor is a lower-level manifestation of the general problem of writing regional geography. As the writing of masters' and doctoral dissertations on regional topics has become unfashionable, so the flow of scholarly regional work into the journals diminished and the textbook writer had fewer well-formed bricks with which to build. This has been somewhat offset by the rising flood of regional statistics but so far these remain undigested and unformed.

Paterson's fifth problem, the submergence of regional distinctions, reflects the diminishing link between localized natural resources and the pattern of human occupation in an area as transport costs fell from the middle of the nineteenth century. This trend undermined the local logic of the vertical structure of man–environment relations in some parts of a regional account: the man–environment links remain but have moved to a global level.

Finally, he suggests that writing regional geography is likely to remain a craft occupation in which the skill and judgement of the individual scholar is at a premium. It is unlikely to have been much influenced by the computer developments mentioned elsewhere in these essays.

Time may also play its part: the timescale of regional change and the timescale of maturing regional geographers. As the pace of technological innovation grows, so regional change is accelerating. But understanding a region involves something much deeper than statistics: it means adopting its culture, learning its language, travelling its byways, scouring its archives, acquiring a specialist understanding of its landscapes and economy. It is costly in time, a business of many years, and not one that fits easily with short project grants or the imperatives of quick results for publication. I can think of none of the regional scholars I have known – men like Bill Mead on Scandinavia, Jim Parsons on Colombia, Harold Brookfield on Melanesia – who had not spent decades in learning their regions in field season after field season.[27]

DECCAN AND GANGES

One of the most complete examples of a regional geography written in Hartshornian terms was provided by Oskar Spate (plate 11) in his *India*.[28] Spate's regional classification of India has three macro-regions – the mountain rim, the Indo-Gangetic plains and the peninsula. These are divided into 34 regions of the first order, 74 of the second order, and about 225 sub-divisions of these (see figure 4.8). The detailed breakdown shows a

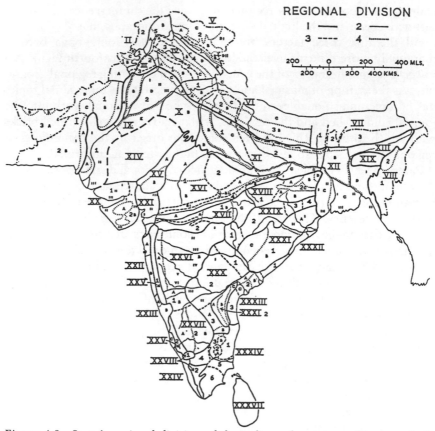

Figure 4.8　Spate's regional division of the Indian sub-continent. The boundaries
are of four weights indicating macroregions (1), regions of the first-order (2), regions
of the second-order (3) and sub-divisions of these (4). See also plate 11.
Source: O. H. K. Spate and A. T. A. Learmonth, *India and Pakistan: A General and Regional
Geography,* 3rd edn. London: Methuen, 1967, fig. 13.1, p. 408.

strong family likeness to the earlier regional schemes of Baker and Stamp,
but a strict delimitation based on one regional master principle is not
attempted.

Structure is perhaps the most obvious guide, at any rate on the
macro-scale; but it is clear that in an area like the Indo-Gangetic Plains
landforms are of little help, since the divisions are as a rule either too
broad or too much a matter of local detail, such as the distinction
between *bet* and *doab, khadar* and *bhangar*; these may be of great
importance in local life, but are in practice useless as bases for regional

XXXIV. TAMILNAD
1. *Coromandel coastal plain:*
    (i) Archean low peneplains: Fs – monadnocks, Cretaceous-Eocene inliers
    (ii) Cuddalore/laterite shelf: Fs – Red Hs, Capper Hs
    (iii) young alluvial zone: Ss – embayments, strandplain; Fs – Korteliyar, Cooum, Adyar, Palar vys
    Madras
2. *Tamilnad Hills (Ø of Mysore Maidan):*
    (*a*) Javadis: Fs – Agaram – Cheyyur through vy, Yelagiri (Ø), Ponnaiyar gap (@ between 1 (i) and 3 (ii))
    (*b*) Sn group: Ss – Shevaroys, Kalroyans, Pachamalais, Salem monadnocks (Ø)
3. *Palar/Ponnaiyar trough:*
    (i) lower shelf of Mysore Ghat (@ to XXVII. 2 (*b*))
    (ii) Baramahal
    (iii) Sn margins (Salem area, @ to 4): Fs – Chalk Hs, magnetite monadnocks
4. *Coimbatore plateau (Kongunad):* Ss – Bhavani, Noyil, Amaravati vys, Palghat sill (@ to XXIV). 1 (ii), Coimbatore Hs; F – Kangayam interfluve
5. *Cauvery delta:*
    (*a*) delta head (@ to 3 and 4)
    (*b*) delta proper:
        (i) Velar/Coleroon doab (@ to 1 (iii))
        (ii) Coleroon/Cauvery doab: Fs – Srirangam Island, floodplains
        (iii) main delta plains: Ss – higher Western margins (F – Vallam Table-land), older irrigated area
        (iv) seaface: Fs – marshy low, dune belt, Pt Calimere, Veddaniyam salt swamp
6. *Dry Southeast:*
    (*a*) upper Vaigai:
        (i) Varushanad vy
        (ii) Kambam vy
        (iii) Dindigul col (@ to 4)
    (*b*) Madurai/Ramanathapuram shelf:
        (i) colluvial piedmont zone: Fs – monadnocks (Sirumalais, etc., Ø)
        (ii) laterite/old alluvium panfan (from Varshalei to Vaippar): Ss – tank country, coastal strip, Pamban Island (Fs – old reefs, Adam's Bridge)
    (*c*) Black Soil area
    (*d*) Tirunelveli:
        (i) colluvial zone
        (ii) red soil zone: Fs – *teris*, coastal dunes
        (iii) Tamprabarni basin: Ss – foothills (@ to XXVIII. 2 (iv)), Chittar vy, Tamprabarni vy

* See figure 4.8. Fs=Features Hs=Hills Ss=Stows Vys=Valleys Ø=Outlier
@=Transitional Sn=Southern
Source: O. H. K. Spate and A. T. A. Learmonth, *India and Pakistan*, 3rd edn. London: Methuen, 1967, table XXXIV, pp. 422–3.

descriptions on a sub-continental scale – and in descriptive writing, scale is of the essence.[29]

Transitional differences in climate may also be used in open plain areas.

On the human side, Spate considers that the best approach in theory might be to relate the regional network to the spheres of influence of towns, but data are lacking and he is dubious of transferring western methods to the Indian scene. Instead, historical identities such as those evoked by the names Tamilnad, Maharashtra and Malwa cannot be ignored. If these traditional regions are not coincident with physical boundaries, neither are they wholly divorced from them. Sometimes a regional boundary will match a longstanding administrative boundary; more often they cut across them. Some boundaries are the mid-points of transitional zones; some are sharply cut. Some regions are rich juxtapositions of nature and human activity overlapping on the map; others, such as the upper Mahanadi basin, are simply what is left over after more definite regions have been sieved out.

For Spate, the essence is understanding rather than some ideal schemata:

> The factors to be considered are far too multiplex to 'drape themselves completely' in accordance with an a priori division, be it on physical or political lines. The *practising* regional geographer cannot bind himself in advance to so rigid a statute; our study is far too rich, varied and subtle to be tied down easily by rule and line.[30]

We can see this pragmatism by looking at his treatment of one of south India's most distinctive regions, the Tamilnad (Region XXXIV on figure 4.8). The Tamil country, now roughly equivalent to the state of Madras, is characterized by two dominant themes: the homogeneity of its Tamil culture and the environmental problems set by its rainfall regime (with most falling in the three months of October to December). But within this uniformity, physical environment is very diverse and the fine tracery of its second-order regions and 25 minor subdivisions is mainly based on geomorphology, soil and climate. But Spate's order is not an imposed or unyielding one; climate enters again to divide the dry Southeast (Region XXXIV–6) from the rest of Tamilnad, and the urban importance of the capital city, Madras, means it must be taken apart from the rest. In its sensitivity to anomaly, transition and outliers, Spate's scheme is a work of craftsmanship in which each piece of wood in the inlaid mosaic is carefully fitted and delicately finished.

HUMANISM AND REGIONAL GEOGRAPHY

For all its power, Spate's *India* belongs to a tradition of regional writing that has attracted relatively few geographers over the last two decades. What has taken its place? One of the strong currents flowing in human geography over the last few decades has been towards humanistic, perceptual studies in which the key roles of individuals and groups within society in interpreting the resources and landscapes of different regions have been stressed. Fisher reminds us that the Swiss historian Jacob Burckhardt regarded history as the study of what the people of one period found significant in another.[31] If we extend the parallel between history and geography, we might define regional geography as 'the study of what the people of one place find significant in another'.

If this interpretation is followed, we should not expect to see a definitive regional geography for each part of the world; rather a series of regional interpretations, each written from a particular perspective. The very success of Walter Prescott Webb's *Great Plains*, a regional monograph by an American historian, was precisely because it allowed the Great Plains environment to be seen successively through the eyes of different genera- tions of settlers. Each generation saw different potentials. The work of interpretation by sensitive regional scholars like Donald Meinig follows in the same vein, allowing the interplay of both the human actors and the environmental stage to show through.[32]

Historical geographer, Sir Clifford Darby, has pointed out the technical difficulties in writing regional geography. He shows that it is a humiliating experience for a geographer to try to describe even a small tract of country in such a way as to convey to the reader a true picture of the reality. Part of the reason for this difficulty is a geometrical one; one-dimensional time means that ten time periods can only be arranged in one of two ways, forwards or backwards. Two-dimensional space means that ten regions can be arranged in several million sequences and it is not obvious which sequence is the best.[33]

This complexity may have something to do with the argument that geographers are intrinsically less able to write well than their historical colleagues. I am not convinced by this argument. Consider, for example, the following passage:

> One afternoon in September, 1918, a British staff automobile left corps headquarters 'somewhere in Flanders', sped eastwards over good roads for a few miles, then plowed into the ruts and mud holes of a newly recaptured portion of His Majesty King Albert's dominions. It was typical Flanders weather, for a drizzling rain was falling and low

clouds or banks of fog drifted over the plain ... Then, as if raised by magic hands, the fog curtain slowly lifted and parted. A flood of golden sunshine burst through, lighting up a vast green-carpeted plain on which rivers and ponds glittered like silver spangles. Stretching in a vast crescent across the stage thus revealed to the waiting observers was a line of flashing tongues of flame, a semicircle of steel and fire, from the sea on the northwest to the uplands of Artois far away to the southward.[34]

There is nothing about the flow of the prose to suggest it was not the work of a war correspondent or military historian. In fact, the author was the American physical geographer, Douglas Johnson, and the passage occurs on the first page of his monograph for the American Geographical Society on *Battlefields of the World War: Western and Southern Fronts* (plate 12). Johnson's narrative runs on easily in the time dimension, but becomes more cumbersome when regional sequences have to be tackled.

GEOPIETY

One of the factors that gives both force to regional writing, yet sometimes clouds its clarity, is a deep love for the landscape being described. Attachment to a particular part of the earth's surface was a favourite theme of the American geographer John K. Wright. The term Wright invented, 'geopiety', has been explored further by Yi-fu Tuan who finds it recurring in all ranges of peoples and at all spatial scales from local to global.[35]

This attachment can range in intensity from a general and intermittent support for a team in the inter-school sports or Olympic Games, to the passionate intensity of the south Tyrolese for his home valleys expressed in the emotionally-charged word Heimat.

When we say the word 'Heimat' then a warm wave passes over our hearts; in all our loneliness we are not completely alone ... Heimat is mother earth. Heimat is landscape we have experienced ... our Heimat is the land which has become fruitful through the sweat of our ancestors. For this Heimat our ancestors have fought and suffered, for this Heimat our fathers have died.[36]

The attachment to land may not be confined to a particular quality of the landscape. There are unlikely devotions to places: those born there cannot bear to leave them; but those not born there could never consider living there.

The same basic sense of identification with the land is a continuing feature which remains strong, even in a society where the urban proportion of the population is steadily growing. The theologian Paul Tillich recalls that:

> Nearly all the great memories and longings of my life are interwoven with landscapes, soil, weather, the fields of grain and the smell of the potato plant in autumn, the shapes of clouds, and with wind, flowers and woods. In all my later travels through Germany and southern and western Europe, the impression of the land remained strong.[37]

Tillich goes on in his *Systematic Theology* to develop the notion of the spirituality of space:

> In reality spirit has its place as well as its time. The space of the creative spirit unites an element of abstract unlimitedness with an element of concrete limitation ... It becomes a space of settlement – a house, a village, a city. It becomes a space of social standing, of community, of work. These spaces are qualitative, lying within the frame of physical space but incapable of being measured by it. And thus the question arises as to how physical space and the space of the spirit are related to each other, i.e. the question of historical space.[38]

In his notion of historical space, Tillich anticipates some of the concern with fossil landscapes which we address at the end of chapter 7. Meantime, it is worth noting that love of region, geopiety, Heimat, all have their reverse sides. Literature is also strewn with the record of man's hatred and fear of particular places.[39] The landscapes of childhood and youth seem to have a particularly strong impact on academic geographers, and they keep returning to that theme in later life: for example, the landscapes of Missouri for Carl Sauer or central Sweden for Torsten Hägerstrand.

QUANTIFICATION AND REGIONAL GEOGRAPHY

The sweep of work in quantification and theory building stands at the other extreme to the humanistic approach to regional geography. Superficially, there is a negative correlation between the rise of so-called quantitative geography in the 1960s with a decline in interest in our first type of regional geography. Most of the mathematical tools developed during the so-called quantitative revolution were appropriate to rather simple geographical systems. By contrast, the notion of a region with its very complex sets of relations between man and land in particular places implies a much more complex system. As John Muir wrote in *My First Summer in the Sierra*, 'When we try to pick out anything by itself, we find it hitched to everything else in the universe.'[40] An example of this interweaving is provided by figure 4.9. The search to tease out some of these complex connections continues to

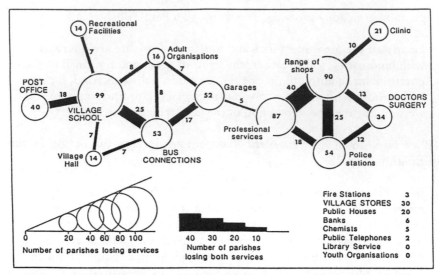

Figure 4.9   Interlinking a complex of regional changes. Links between the closure of village schools and loss of other services in an English rural county over a thirty-year period. The width of the bonds is proportional to the number of parishes losing a pair of services.

Source: P. Haggett, E. A. Mills and M. A. Morgan, *An Atlas of Rural Services in Somerset and South Avon, 1950–1980: A Preliminary Analysis.* Bristol: Final Report to the Social Science Research Council, Vol. 6 (1983), pp. 1–44; fig. 6, p. 43.

be at the heart of regional analysis. One of the critical reasons for the relative neglect of regional geography may therefore be purely technical; it was ignored not because it was unimportant but because it lay beyond the capability of a new generation of scholars armed with powerful but essentially crude analytical tools.

In retrospect, I suppose more could have been done with the first generation of quantitative models than we realized at the time. Suppose that we had wished to adopt three guidelines in our regional analysis: to tune the scale of our regions to the dominant 'wavelength' of the spatial process being studied; to restrict our discussion to only the most critical ecological links treated in some optimal order; to tailor the complexity of our regional jigsaw to a relevant teaching level.

Light may be thrown on each of these by critical applications of an appropriate analytical method. Spatial autocorrelation models help in the first analysis, systems dynamics in the second, discriminant analysis in the third.

Let me illustrate the point most simply by the third case. The Swiss geographer, Dieter Steiner, developed a quantitative taxonomy for the climatic regions of the United States. This measured precisely the loss of

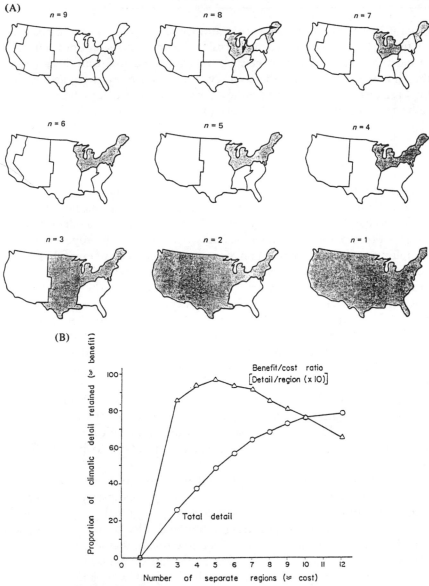

Figure 4.10   Trade-off between information and complexity in regional taxonomy.
(A) Steiner's optimum climatic regionalization of the United States based on nine
regions (*n*=9) through one region (*n*=1). (B) Impact on the level of climatic detail
with changes in the number of regions. The five-region solution appears to be the
most efficient.

Source: Data from Dieter Steiner, 'A multivariate statistical approach to climatic regionalization and classification'. *Tijdschrift van het Koninklijk Nederlandsch Aardrijkskrundig
Genootschap Tweede Reeks*, Vol. 82 (1965), pp. 329–47. Redrawn for Peter Haggett, 'Future
trends in regional geography'. In P. C. Forer (ed.), *Futures in Human Geography*. Christchurch: New Zealand Geographical Society, 1980, fig. 1.1, p. 9.

generality (or gain in detail) as the number of climatic regions increased.[41] As figure 4.10 illustrates, each successive regional division does not bring proportionate benefits: some new boundaries bring important insights into the climatic variability of the country, others are more marginal. A 'best buy' regionalization of United States climate on the Steiner scheme appears to be at the third split where five separate regions are recognized.

NEW TRENDS IN REGIONAL GEOGRAPHY

It is important to recognize that the quantitative movement in human geography has itself been undergoing change. To the earlier statistical traditions has been added a more broadly based interest in modelling which stresses other parts of mathematics and is better able to cope with more complex non-linear systems. Penn State geographer, Peter Gould, has used a new type of algebra ('polyhedral' algebra) to model flows in television programmes between one world region and another. The export of the BBC's *All Creatures Great and Small* to Scandinavia represents one such flow, the import of Australia's *Neighbours* into the UK another. The battle for minds and regions takes a new form in the late twentieth century, with TV replacing gunboats. Forrester's *World Dynamics* has, despite its own considerable limitation, sparked off a new wave of models which simulate global and regional futures.[42]

It is an optimistic but not unreasonable scenario to see a blending of the two separate traditions of regional geography and regional science. Regional geography could benefit from some additional analytical tools to add to its broad-gauge cogitations; regional science needs to go beyond formal elegance.

So there are reasons for some optimism on the relative importance of regional geography within the university sector if one is prepared to accept a catholic definition for the field. I would expect regional geography of the classic kind to become more interpretative and selective, or possibly more varied and more controversial. The reference role of regional geographies is increasingly being served by the fine collection of national and regional atlases that has long been appearing. If designed with imagination and written with full critical commentary, these provide the most useful of sources, both as a review of what has been achieved and a basis for further research. Such atlases as Schwartzberg's *India* or Duncan's *Victoria* are superb examples of work in this genre.[43] The integration of regional geography with quantitative model building is proceeding more slowly, both for technical reasons and for the lack of overlap in interests. But geographers are now learning to swim in the rising tide of data, rather than drowning in numbers.

5

The Arrows of Space

On second thoughts she made up her mind to go on: 'for I certainly won't go back,' she thought to herself, as this was the only way to the Eighth Square.
Lewis Carroll, *Through the Looking Glass and What Alice Found There* (1872)

One of the delights of a life as a geographer is that, as a profession or a passion, it provides a copper-bottomed excuse to spend time in the oddest corners of the world. I was appreciating this compensation (for, as my family will confirm, there are down sides too) when I spent November 1984 in the Fijian Islands. They consist of two main islands, Viti Levu and Vanua Levu, and hundreds of smaller islands, atolls and reefs scattered over an ocean area the size of France. The islands of the Pacific have had a fascination for scientists that goes well back beyond Charles Darwin and his evolutionary study of the Galapagos finches. For human geographers, the discovery and settlement of the islands, the build-up of their population, and development of distinctive cultural and economic regions in Micronesia, Melanesia and Polynesia have all drawn a following.

My own reason for being in Fiji was less romantic. I was in the middle of a study of the way in which western contacts had brought disease into the native populations of some of the islands and the disastrous demographic impact that had sometimes resulted.[1] Fiji is an especially interesting case since it involves the introduction and rapid transmission of an infectious viral disease, measles, which although relatively mild and a disease of childhood in western countries takes on a different cast in previously uncontacted, virgin-soil populations. In January 1875 a British ship, *HMS Dido*, had arrived at the Fijian capital of Levuka bringing back a party of royal chiefs from a state visit to the Governor of New South Wales at Sydney (see plate 13). One of the Fijian princes was recovering from the disease but was still infectious, and the measles virus was evidently passed on to members of the party that came out to meet them. The result was to be a tragedy. Within five months more than 28,000 of Fiji's then population of 150,000 had died directly or indirectly from the disease.

(A)

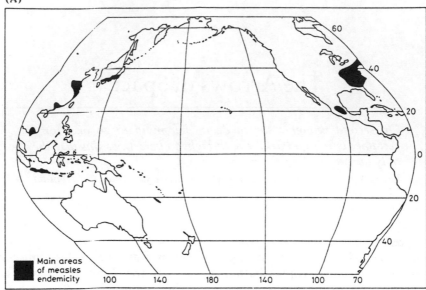

(C)

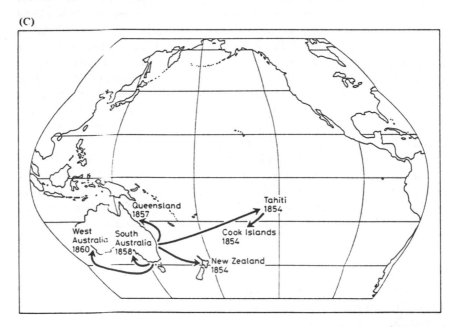

Figure 5.1 Vector maps of a diffusion process. Vectors showing the general directions of measles spread in the Pacific Basin between 1800 and 1910.

Source: A. D. Cliff and P. Haggett, *The Spread of Measles in Fiji and the Pacific: Spatial*

(B)

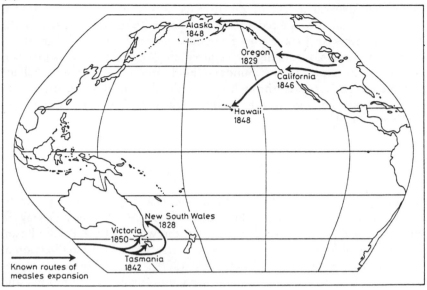

(D)

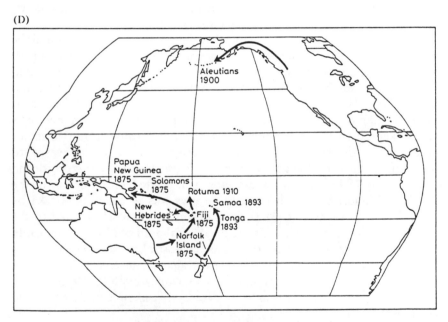

Components in the Transmission of Epidemic Waves through Island Communities. Canberra: Australian National University (Research School of Pacific Studies, Department of Human Geography, Publication HG. 18, 1985, fig. 3, pp. 22–3.

While the general extent of the disaster was well known, the spatial pattern could only be pieced together from fragmentary sources in the national archives at Suva. I found the most useful from the geographical viewpoint the records from my own church, the Wesleyan. Always prone to counting souls and keeping Sunday-school and class registers, village mission churches scattered throughout the island chains had records which confirmed both the scale and the swiftness of what *The Times* was to call 'the Fijian tragedy'.[2]

VECTORS AND FATAL SHORES

The single Fijian event takes on spatial significance only when it can be placed in a much broader hemispheric context. Figure 5.1 shows the ways in which the same disease spread throughout the whole of the Pacific Basin. The sequence of four maps shows, using vectors, the generalized direction of measles spread in the Pacific Basin since 1800.[3] The technique of vector mapping is straightforward. Dates of known occurrences are plotted and the likely corridors of movement between places are linked using vectors, with the arrowheads denoting the direction of spread.

Map A illustrates the most probable distribution of the disease around the rim of the Pacific Basin at the start of the nineteenth century. It was then endemic in the high-density population clusters of south and east Asia, and was already established in eastern North America, central Mexico, Peru and central Chile. The spread during the first half of the nineteenth century is shown in map B. During this period, the 'virgin soil' state of many areas was broken down. As the vectors show, the attack came from two major directions. In the northern Pacific, the disease followed the westward movement of the colonists across North America. Measles was reccorded in California in 1846 and two years later had started an epidemic in the Hawaiian Islands. Still further north, measles was recorded in Alaska by 1848. The Californian and Alaskan routes are also closely related to the great gold rushes of the period on the Pacific seaboard of North America. In the southern Pacific, the main focus was on Australia. Newspaper records suggest that measles was unknown in Australia for the first forty years of British settlement, until the first outbreak in Sydney in 1828. Despite strict quarantine regulations, measles reappeared in Tasmania in 1842 and in Melbourne in 1850.

Map C shows the spread vectors during the critical decade from 1851 to 1860. The process of diffusion appears to have gone through two distinct phases. First, spread occurred from Victoria and New South Wales to the other Australian states; Queensland (1857), South Australia (1858) and

West Australia (1860). In all three instances, the virus appears to have been carried by local shipping contacts between the Australian colonies, rather than to have been introduced directly from Europe. The second phase saw the introduction of measles in 1854 into New Zealand, Tahiti and the Cook Islands.

The final map, D, illustrates the closing stages of the invasion process from 1861 to 1910, by which date the present distribution of the virus in the Pacific had been largely established. The year 1875 was one of the most significant in the history of measles spread in the region since a major outbreak in Sydney was carried to a number of neighbouring islands in the southwest Pacific. These included the disastrous Fiji episode in 1875 and the later spread to Tonga and Samoa (1893). Meantime in the north Pacific measles was introduced into the Aleutians in 1900.

The maps show that the spread of measles into the island populations of the Pacific appears to have been a nineteenth-century phenomenon. In 1800, measles seems to have been unknown in the basin; by 1900, it had visited most of the main island groups at least once. In this century, the increasing volume of contacts and decreasing travel times allowed the virus to be transmitted ever more readily, particularly once the reservoir areas for measles had become established on the Pacific margins in the south-west (Australia and New Zealand) and in western North America, as well as its traditional hearths in South and East Asia.

FROM VECTORS TO WAVEFRONTS

I have used the Pacific example to show that, at one geographical scale, individual vectors of spread can often be identified. But at another, more aggregated level it may be more useful to think of a spread across as a continuous wave. This switch from individual quanta (the discrete vector) to continuous waves is one that is familiar in physics. The most well-known work on diffusion processes as waves was carried out by the Swedish geographer, Torsten Hägerstrand, who was for many years professor at Lund University.[4]

Hägerstrand selected the Asby district of south-central Sweden to study the spatial pattern of acceptance by local farmers of various agricultural innovations (see plate 14). One such was a subsidy which the Swedish government granted from 1928 onwards to small farmers (farming up to eight hectares of tilled land) if they enclosed woodland on their farms and converted it to pasture. The study area was divided into more than one hundred 5 km×5 km cells (see figure 5.2). For each cell, the total number of farms which had accepted the subsidy by the ends of each of the four years

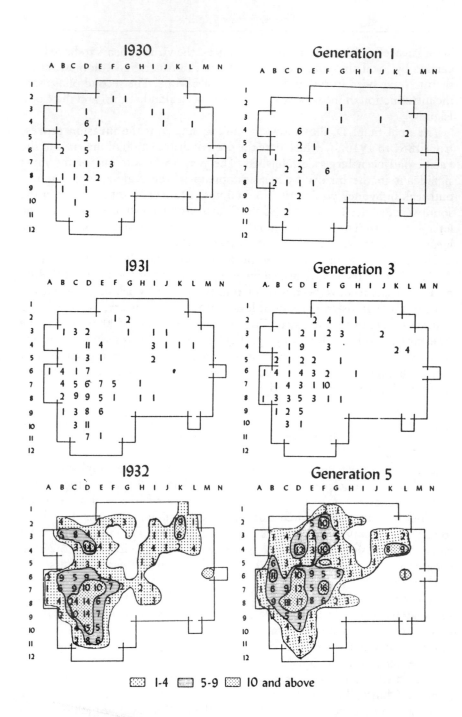

from 1929 to 1932 was recorded. Hägerstrand then developed a diffusion model to simulate the spread of the subsidy through the farms up to the end of each year.

The rules for the model were very simple. Hägerstrand assumed that the decision of a potential adopter (any small farmer) to accept the subsidy was based solely upon information received directly from other farmers at face-to-face meetings. Hägerstrand had noted that the spatial development of many diffusion processes seemed to be by the accretion of new adopters around the original nuclei of introduction; this contagious growth has been called the 'neighbourhood effect'. To take this into account, the model assumed that the probability of a potential adopter being paired with a carrier had a strong inverse relationship with geographical distance between the teller and the receiver.

The formal structure of the model is described in the standard textbooks[5] but four general points are worth noting. First, the numbers and spatial locations of adopters put into the model were the actual configurations at the start of the diffusion process. Second, a potential adopter was assumed to accept the innovation as soon as he was told by another adopter. Third, at each cycle of the model, every adopter was allowed to contact one other person, adopter or non-adopter. Fourth, the probability that a carrier in one cell would contact an individual located in another cell of the study lattice was governed by a floating 25-cell grid, which Hägerstrand called a 'mean information field' (MIF).

Probabilities in each cell of the MIF were estimated from an analysis of migration and telephone traffic data in central Sweden. Probability of adoption was highest over the central cell of the MIF and declined exponentially to become zero at the most distant cell, eighteen kilometres away. The floating grid was placed over each existing adopter cell in turn, so that the adopter was located in the central cell of the MIF.

To take into account the reduction in communication between small farmers likely to be caused by terrain (e.g. access would be interrupted by rivers and lakes), simple barriers were introduced into the model plane. For

---

Figure 5.2   Real and simulated diffusion patterns. Simulation of the spatial pattern of the number of adopters of improved pasture subsidy in central Sweden. From the initial distribution of adopting farmers in 1929 (Generation 0) the next three years are shown; the actual distribution on the left (1930, 1931, 1932) and the simulated distribution on the right (Generations 1, 2 and 3). Each square cell on the map is 5 km×5 km and numbers give the number of adopters in each cell.

Source: T. Hägerstrand, 'On Monte Carlo simulation of diffusion', *Northwestern University, Studies in Geography*, Vol. 13 (1967), fig. 3, p. 23; redrawn by M. H. Yeates in *An Introduction to Quantitative Analysis in Human Geography*. New York: McGraw Hill, 1974.

example, when a potential adoption crossed a 'half-contact' barrier, the adoption was cancelled with probability of one half. However, two contacts across this barrier allowed spread into the cell. Other barriers to spread could be made more or less permeable.

Using this model, Hägerstrand performed a series of computer runs to simulate the spatial pattern of acceptance of a subsidy for the improvement of pasture on small farms in the Asby district of central Sweden (see figure 5.2). Hägerstrand gives the results from three runs of the model, and we reproduce one of these outcomes. Note the spatially highly contagious build-up of adopters is restricted by the model, something which we would expect on both practical and theoretical grounds.

FROM FORECAST TO CONTROL

Hägerstrand's pioneering work was essentially concerned with simulating spatial processes, i.e. providing a well-defined copy or mimic of an observed historical sequence. In the years following its publication in 1953, his model was adapted to apply to regional cases from the spread of the black ghetto in Seattle to the settlement of Polynesia.[6] The Polynesian case is referred to in the last essay (see figure 8.4).

But what use is this spatial model? To see this, we have to look at the way in which science proceeds. Given the diffusion process being studied, there are three relevant questions. First, what is happening? From an accurate observation we may be able to answer this and summarize our findings in terms of a descriptive model (like the Hägerstrand models). Second, what will happen in the future? If our model can simulate past conditions accurately, then we may be able to go on to say something about future conditions. This move from the known to the unknown is characteristic of a forecasting model. The basic idea is summarized in figure 5.3. But planners and decision-makers may want to alter that future, say, to accelerate or stop a diffusion wave. So our third question is: What will happen in the future, if we intervene in some specified way? Models which try to accommodate this third order of complexity are called planning or intervention models.[7]

I was fortunate in the late 1960s to have a Canadian doctoral student at Bristol, Roly Tinline, who was interested in extending the Hägerstrand model to this third stage of control situations.[8] An opportunity arose in the aftermath of the great 1967–8 epizootic of foot-and-mouth disease (FMD) in Britain. FMD is a virus disease of cloven-footed animals (cattle, sheep, pigs and goats). Since 1892, control in the United Kingdom has been exercised through total slaughter of all exposed herds, because the virus is

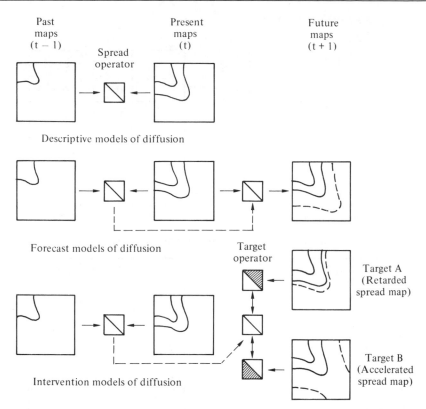

Figure 5.3 Map sequences as a forecasting device. Descriptive, forecast and intervention models of diffusion. The creation of mean information fields (MIFs) from a map sequence allows the forward projection of the map to a future time period.

so readily transmitted among cattle. Even if an animal survives, the economic value of an infected animal is much reduced. The aim of this policy is to destroy all possible reservoirs of infection and thereby eliminate the disease.

Tinline explored the use of geographically targeted vaccination strategies, in conjunction with slaughter, as a more cost-effective way of containing FMD. He was able to demonstrate, using data from the outbreak, that airborne spread of the virus downwind from sources of infection was a major cause of additional outbreaks. By the time the disease was positively identified in a core area and slaughter carried out, virus particles had often been carried long distances by the wind. These virus particles were redeposited after rain in areas well beyond the FMD slaughter zone. The

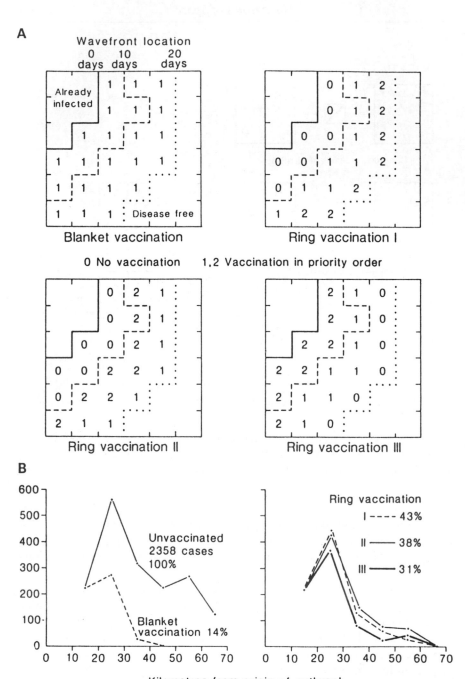

A

Wavefront location
0   10   20
days days days

Already infected

Blanket vaccination

Ring vaccination I

0 No vaccination   1,2 Vaccination in priority order

Ring vaccination II

Ring vaccination III

Disease free

B

Unvaccinated
2358 cases
100%

Blanket
vaccination 14%

Ring vaccination
I ---- 43%
II —— 38%
III —— 31%

Kilometres from origin of outbreak

shape of the mean information field identified by Tinline for FMD was different from that of Hägerstrand (see figure 5.4). Its assymetric structure reflects the wind-blown form of spread with direction as well as distance from the central cell affecting the probability of spread.

To overcome the rapid spread characteristics of FMD, Tinline tested different schemes of ring vaccination in areas downwind of initial outbreaks to try to contain subsequent spread. The principles involved are illustrated in figure 5.4. Ideally, immediate vaccination of all herds downwind of the outbreak is needed. However, this blanket strategy is generally impossible in any major outbreak because of shortage of vaccination teams. Using data for the outbreak, Tinline therefore simulated the impact of different geographic arrangements of vaccination throughout areas downwind of any outbreak. He found the response curve to increasing vaccination levels to be one of diminishing returns, with vaccination out to a distance of 20 km from an outbreak being necessary to construct a 'cordon sanitaire' to contain the disease. Of the three strategies shown in figure 5.4, that of Model III with vaccination proceeding from the margins inwards proved the most effective.

The practical difficulty in implementing a ring vaccination policy is the need to forecast accurately wind conditions in the vicinity of an outbreak; these will dictate both the location and geographical extent of the ring. The presence of instabilities in the air currents (e.g. lee waves) will further complicate affairs. Given the need to mobilize vaccination teams and the length of time required for vaccination to induce immunity in animals, wind forecasts for periods several days ahead were required for optimal results. This remains the main limitation on applying the model in field situations.

MULTIPLE WAVE MODELS

So far in this chapter we have been concerned with a very simple diffusion process: a single pulse which starts in a limited part of the earth's surface and spreads out quickly or slowly to a larger area. We can envisage this as a stone tossed into a pool with the ripples spreading out, ring-like around the

---

Figure 5.4 Spatial control strategies for countering disease spread. Simulated effect of blanket vaccination and ring-control strategies on the spread of the 1967–68 foot-and-mouth epizootic in central England. Each cell on the maps in (A) covers a 10 km × 10 km area.
Source: Rowland Tinline, A Simulation Study of the 1967–8 Foot-and-Mouth Epizootic in Great Britain. Doctoral dissertation, University of Bristol, England, 1972, fig. 6.17, p. 300 and table 6.12, p. 303.

point of impact. Since in geographical situations the medium through which real diffusion processes are operating is much more complex than water, the waves are not smooth, but otherwise the analogy is reasonable.[9]

Interest in wave-generating processes by mathematicians goes back at least to Isaac Newton's interest in tides. Since then climatic cycles, economic cycles and demographic cycles have also attracted mathematical interest. In medicine, the regular alternation of epidemic and non-epidemic years in such infectious diseases as measles attracted the interest of the Oxford statistician, Maurice Bartlett,[10] and his findings are illustrated in figure 5.5.

This shows three communities of decreasing population size (A, B, and C), each with characteristic epidemic waves (Types I, II, and III). Communities in the first category were large enough to maintain a permanent reservoir of a given virus. Bartlett's study of British and American cities suggested that between four and five thousand measles cases each year in a community were just enough to sustain such a reservoir. Given the reporting and attack rates prevalent at the time of Bartlett's study in the 1950s, this implies that for the measles virus a city of around a quarter of a million inhabitants is the lower population size threshold for a permanent reservoir to exist.

What happens below this level is shown in figure 5.5. In large cities above the size threshold, like community A, a continuous trickle of cases is reported. These provide the reservoir of infection which sparks a major epidemic when the population at risk (the susceptibles) builds up to a critical level. Since measles confers subsequent lifelong immunity to the disease, this build-up occurs only as children are born, lose their mother-conferred immunity and escape vaccination or the disease. Eventually the susceptible population will become sufficiently large for an epidemic to break out. When this happens, the susceptibles are diminished and the stock of infectives increases as individuals are transferred from the susceptible to the infective population. This generates the characteristic 'D'-shaped relationship over time between the sizes of the two populations shown on the end plane of the block diagram.

If the total population of a community falls below the quarter of a million size threshold, as in settlements B and C, measles epidemics can only arise when the virus is introduced into it by the influx of infected individuals (so-called 'index' cases) from reservoir areas. These movements are shown by the broad arrows in the diagram. In such smaller communities, the susceptible population is insufficient to maintain an unbroken chain of infection. The disease dies out and the susceptible population grows unchecked in the absence of infection. Eventually it will become big enough to sustain an epidemic when another index case arrives. Given that the total population of the community is insufficient to renew by births the

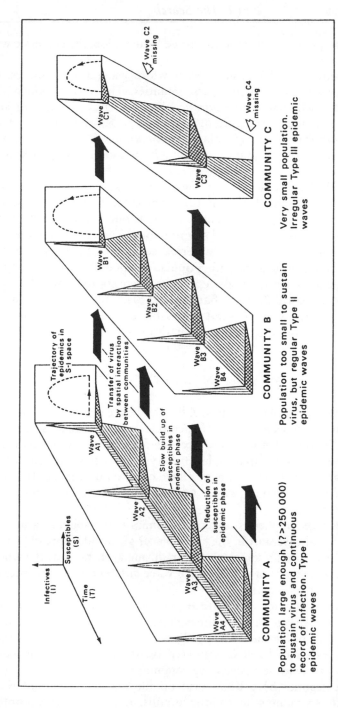

Figure 5.5 Bartlett wave model of disease spread. Measles waves transmitted through geographically separated communities of different population sizes. The diagrams describe the situation in a sequence of communities with diminishing population size: see description in text.

Source: A. D. Cliff and P. Haggett, *Atlas of Disease Distributions.* Oxford: Basil Blackwell, 1988, fig. 6.5(A), p. 246.

susceptibles as rapidly as they are diminished by infection, the epidemic will eventually die out.

It is the repetition of this basic process which generates the successive epidemic waves witnessed in most communities. The way in which the continuous and regular epidemic waves of A break down, as population size diminishes, is emphasized in figure 5.5. Very large but irregularly-spaced epidemics can occur in small and isolated communities (C) if build-up takes place in the absence of vaccination and index cases over a long period of time.

Bartlett's calculations for cities were re-examined by the Yale epidemiologist, Francis Black.[11] This time, however, the units of observation were not cities but eighteen islands. Of these, only Hawaii with a total population of over half a million had a continuous Type I pattern of measles infection over the sixteen-year study period considered by Black. Type II patterns were found in islands with population totals as low as 10,000. Below this population level, measles epidemics were sporadic and the Type III pattern predominated. Both the Bartlett and Black results show how the wave-making engines work to send trains of disease pulses around the world. We now turn to the spatial order in those waves.

SPATIAL ORDER IN MULTIPLE WAVES

For much of the last ten years, Andrew Cliff and I have made a special study of the epidemic waves which have affected one of Black's islands, Iceland, located in the North Atlantic.[12] The reasons for choosing this small and remote country as a geographical laboratory were set out in an earlier chapter (see page 73). For the moment all we need to know is that between 1945 and 1970 eight distinct waves of measles affected the children of Iceland.

What interests geographers studying epidemics is the spatial pattern of the spread. Did it always start in the same place? How did it spread? Did it move more slowly through some areas than through others? Was the whole island affected or just part of it? Was each wave different in its incidence or was there some common pattern?

The four maps in figure 5.6 illustrate some of the answers. This is a lag map showing average lags based on the eight episodes. It was constructed by measuring for each of Iceland's fifty medical districts the length of the time gap between when measles first appeared on the island in a particular epidemic wave and when it reached the community in question.

Four time categories have been used on the maps: fewer than three months, three to six months, six to nine months, and nine or more months.

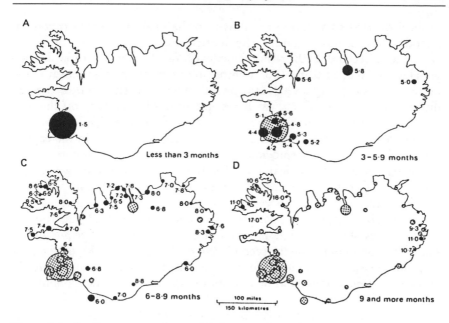

Figure 5.6   Stable patterns of spread behaviour. Mean lag time in months from the start of all measles epidemics occurring in Iceland between 1945 and 1970 until medical districts were reached. Circles are proportional to district populations in 1970. Those in black show districts reached in the time band shown; those reached in earlier time bands are stippled.

Source: A. D. Cliff, P. Haggett, J. K. Ord and G. R. Versey, *Spatial Diffusion: An Historical Geography of Epidemics in an Island Community.* Cambridge: Cambridge University Press, 1981, fig. 5.2, p. 69.

The circles are proportional to the population size of the districts in 1970. Solid circles indicate that a district was first reached in the period covered by the map; stippled circles are used to denote districts shown as reached on an earlier map.

The first map, A, implies that the national capital, Reykjavik (then with a population of 73,000),[13] is generally the first district affected in each epidemic, on average within six weeks of the start. In the second time period, three to six months, two spatial processes are evident, namely, intensification of the epidemic in the districts immediately adjacent to Reykjavik, and long-distance spread to regional centres in the north and east. Akureyri (population 10,000) and Egilsstadir (population 1,700) are the main shopping centres in the northern and eastern parts of the country. The third map, covering the period from six to nine months, sees the disease reaching almost all medical districts around the coast of Iceland. The last medical districts to succumb are shown in the final map which covers the period of nine or more months. The districts concerned are concentrated in two areas of Iceland. In the remote northwest fjords, there are three districts with lags of 10.6, 11.0 and 16.0 months. In the eastern fjords there are three more districts with lags of 9.3, 10.7 and 11.0 months. The remotest medical district, the island of Flatey (with a population of only 182) off the west coast, is the last affected with an average lag of 17 months. Thus on average it is nearly a year and a half after a measles epidemic starts in the capital, Reykjavik, that the same epidemic reaches the remotest community.

What kind of geographical spread pattern do the maps show? At first inspection the patterns described support the idea of a spread model for measles epidemics which proceeds from large places to small (that is, down the urban size hierarchy), associated with contagious spread out from the initial centres of introduction into their hinterlands. The average population size of the communities on the second map of B is nearly 4,000; on the third map the average population size is 1,400; on the fourth map, where the average lag-time is longest at nine or more months, the average population size is less than 900.

This simple interpretation is unfortunately complicated by the fact that the population size of settlements in Iceland generally decreases with distance from both the capital (Reykjavik) and from the regional centres (Akureyri, Egilsstadir and Isafjordur). In order to disentangle the effects of size and distance, various statistical techniques can be employed,[14] which confirm that average time to infection increases as population size decreases and that it also increases with distance from Reykjavik. As is implied by the map sequence, when medical districts become smaller and more remote, so they are reached later and later in the history of any epidemic. Of the two effects, population size appears to be a more important determinant than distance. Geographers term the first process 'cascade diffusion' and the second 'wave diffusion'. 'Cascade' implies a process splashing step by step down the urban hierarchy, while 'wave' describes a local spreading process.

RECONSTRUCTING GHOST WAVES

In the case of Iceland, as in our earlier Pacific example, the process of spread is clear. By searching the historical record and marking on the map the origin of the infection, the route followed by the index case and the place of arrival, a series of vectors can be drawn on the map with some certainty. But reconstruction is rarely that straightforward. More commonly, a disease will break out in a particular location but the origin remains a puzzle. Just how it arrived there and from where is not known.

PLATE 1 Landscape in a mirror. The terraces of the Shotover River, New Zealand (compare with fig. 1.1) are cut in the gravel filling of a former valley while the river has been excavating new gorges through buried spurs on which it is superimposed. The pattern of terraces has been somewhat disfigured by excavations made by sluicing gold-bearing gravel. The road up the valley is unsafe in parts and when I followed it had a notice telling drivers, 'car insurance is invalid past this point'. (Source: Photography by V. C. Browne, in C. A. Cotton, *Geomorphology: An Introduction to the Study of Landforms.* Christchurch: Whitcombe and Tombs, 1942, fig. 250, p. 247)

PLATE 2 Geographers I. Carl Ortwin Sauer (1889-1975), member of the influential Berkeley department of geography for fifty years, was one of the leading influences on human geography this century. He is seen here photographed by Karl Pelzer in September 1935. (Source: Frontispiece to John Leighley, editor, *Land and Life: A Selection from the Writings of Carl Ortwin Sauer.* Berkeley: University of California Press, 1963)

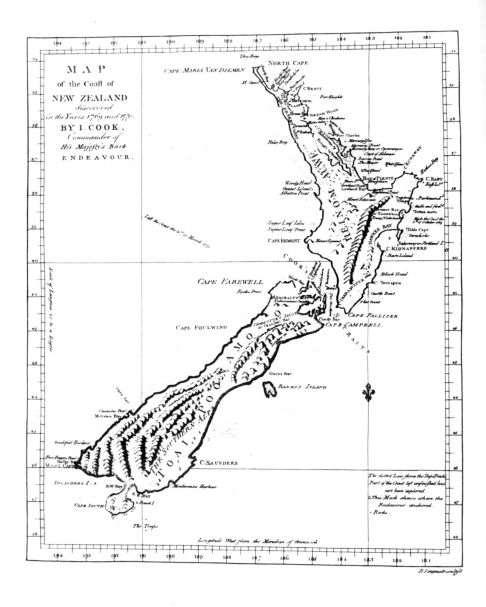

PLATE 3 Exploration. 'A Chart of New Zealand or the Islands of Aeheinomouwe and Tovypoenammu lying in the South Sea. By Lieut J. Cook, Commander of the *Endeavour*, 1770.' See discussion on pages 28-31. (Source: British Museum, Cook, Additional Manuscript 7085, folio 17. This later version is from the Canterbury Museum, Christchurch)

PLATE 4   Geographical sampling. Northern part of the Fortaleza basin, Taubate basin, Brazil, showing the location of some of the 256 sample points on which a factorial analysis of land use was based. Compare with figs 2.5 and 4.4. (Source: Peter Haggett, 'Regional and local components in the distribution of forested areas in southeast Brazil: a multivariate approach', *Geographical Journal*, vol. 130 (1961), pp. 50-9, Plate 2, facing p. 369)

PLATE 5   Decoding maps. The much debated question of the existence or non-existence of so-called 'clusters' of childhood leukaemia cases near nuclear power stations has attracted geographical work. See the discussion on pages 35-8. The nuclear power station at Hinkley Point. West Somerset, is shown here. (Source: Hunting Aerofilms Ltd)

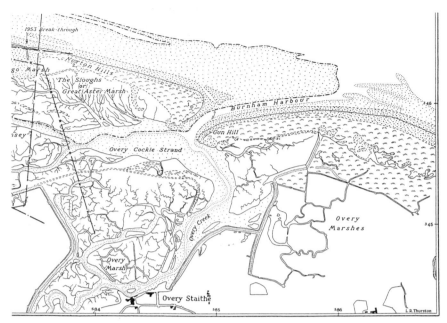

PLATE 6 Field mapping I. Section from a map of changes in the shingle bars, sand dunes and marshes of Scolt Head Island off the north coast of the county of Norfolk, East Anglia, England, as surveyed by two undergraduates (the author with John Small) under the direction of A. T. Grove and J. A. Steers, long vacation 1953. See the discussion on page 46. (Source: J. A. Steers, ed., Scolt Head Island, Cambridge: Heffer, 1960, endpiece)

PLATE 7 Field laboratories I. The scattered population of hamlets and villages in Iceland have provided a valuable field laboratory for tracking the geographical spread of childhood diseases. See the discussion on pages 73-5. Skutustadir, Lake Myvatn, northern Iceland.

PLATE 8 Field mapping II. Vila Verde, a small irrigated property in the Ancora valley of northwest Portugal. Compare with the field map in fig. 3.1(B). (Source: Photograph by author in R. D. Hayes, 'A peasant economy in north-west Portugal'. *Geographical Journal*, vol. 122 (1956), pp. 54-70, fig. 7, p. 56.)

D

1904

PLATE 9   Centres of geographical spread. (A-D above and opposite) The Lutheran church at Eyri, the focal point of an epidemic outbreak in northwest Iceland in May 1904. (A) General view showing the location of the church on a single pit jutting out into the fjord. (B) The church looking across the main fjord. (C) Church interior. (D) Entry from church register for May and June 1904 showing measles deaths. See the discussion on pages 75-8. (Source: A. D. Cliff, P. Haggett and R. Graham, 'Reconstruction of diffusion processes at different geographical scales: the 1904 measles epidemic in northwest Iceland. *Journal of Historical Geographical*, vol. 9 (1983), pp. 29-46, fig. 8, p. 39)

PLATE 10   The impact of geographical separation. The Tasman bridge linking the two sides of the city of Hobart, Australia, was smashed by an ore-carrying ship in January 1975. Smashing this vital link posed east-side car commuters with a thirty-mile additional drive or long queues for ferry boats. (Source: photograph by the author)

PLATE 11 Geographers II. Oskar Spate, the dean of Australian geographers, was foundation professor at the Australian National University and Director of the Research School of Pacific Studies. He is best known for major works on the geography of the Indian subcontinent and the Pacific basin. Photo courtesy of Professor R. G. Ward, Australian National University.

PLATE 12 Field descriptions. Block diagram by F. K. Morris of the Chemin des Dames battlefield in northern France showing the salient features of the terrain. The first battle of the Chemin des Dames began in April 1917. (Source: Douglas Wilson Johnson, *Battlefields of the World War: Western and Southern Fronts. A Study in Military Geography.* New York: Oxford University Press, 1921 (American Geographical Society, Research Series, No. 3), fig. 64, p. 234).

PLATE 13 Geographical diffusion. A critical vector in the spatial spread of measles in the south Pacific (compare with fig. 5.1) was provided by the voyage of HMS *Dido* in January 1875. See the discussion on page 95. The photograph shows the ship at Portsmouth in 1871 or 1876 and is from the records of the National Maritime Museum, Greenwich. It is reproduced in A. D. Cliff and P. Haggett, *Atlas of Disease Distributions,* Oxford: Basil Blackwell, 1988; fig. 5.4, p. 179.

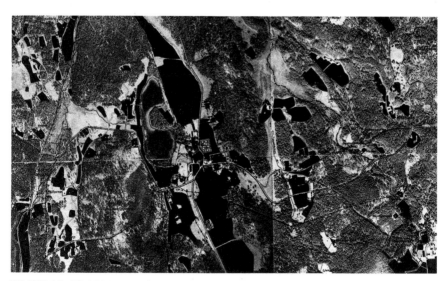

PLATE 14 Field laboratories II. The village of Asby in the wooded area of central Sweden provided data for the classic early migration studies by Torsten Hägerstrand. The 1 : 10,000 map shows cultivated fields in black. (Source: Ekonomisk Karte over Sverige 1 : 10,000 photographs 1945, with additional cartographic detail, 1948)

PLATE 15 Family history. The silver plate given to Alfred Steers on his retirement from St Catharine's College, Cambridge, by geographers from the college now holding posts around the university world. See discussion on pages 121-2. Reproduced by courtesy of Dr James Steers.

PLATE 16   Geographers III. Torsten Hägerstrand, professor of geography at the University of Lund, pioneered many developments in human geography, including spatial diffusion studies. He is here receiving an Honorary Doctorate from the Nobel Prizewinner, Dorothy Hodgkin, Chancellor of the University of Bristol, July 1974.

PLATE 17   Geographical societies I. The main hall of Lowther Lodge, the Hyde Park home of the Royal Geographical Society in London. Founded in 1830, the society has been closely identified both with world exploration and the foundation of geography in British universities. (Source: Royal Geographical Society)

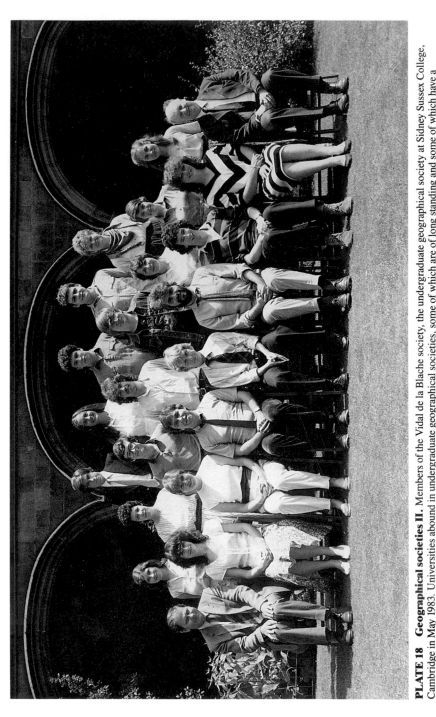

PLATE 18 Geographical societies II. Members of the Vidal de la Blache society, the undergraduate geographical society at Sidney Sussex College, Cambridge in May 1983. Universities abound in undergraduate geographical societies, some of which are of long standing and some of which have a roller-coaster existence, depending on the enthusiasm of the group in residence at any one time. The more careworn looking members in the front row whose undergraduate days are now some way behind them include (from left to right) the author, Dr Derek Gregory, Dr Graham Smith and Professor Richard Chorley. See discussion on page 131.

PLATE 19   Geography departments. (Above) The department of geography at the
University of Canterbury, Ilam campus, Christchurch, New Zealand. The original department
was established in 1937. (Below) Academic staff at one of the world's newest geography
departments at the University of Brunei Darruslam, at Bandar Seri Begawan, a small city on the
east coast of Borneo. The first graduates took their degrees in September 1989. The wide spread
of world geography is emphasized by the fact that staff hold degrees from thirteen universities in
eight countries: Australia, Canada, Hungary, Japan, Malaysia, Sri Lanka, United Kingdom and
the United States. (Photographs by the author)

PLATE 20 International networks. The sixth of the biennial European Colloquia on Theoretical and Quantitative Geography, was held at the Château Les Fontaines, Chantilly, September 1989. Fifteen European and several non-European countries were represented at this meeting of geographers with co-workers in related mathematical fields.

DRAWING BY ALAIN; © 1953
THE NEW YORKER MAGAZINE, INC.

"Want to know something, Dad?"

PLATE 21 Paradigm shifts. Cartoon drawing by Alain; © 1953, 1981, The New Yorker Magazine, Inc. See the discussion on page 143.

PLATE 22   Geographers IV. Griffith Taylor (1888-1963), the controversial Australian geographer who made vigorous contributions to the debate over environmental determinism. Originally a member of the Scott polar expedition, his later work on the limitations of the arid lands in Australia made him unpopular in Australian government circles in the 1920s. His farsightedness was only belatedly recognized as in this 1978 postage stamp.

PLATE 23   Electronic revolution. The impact of the electronic revolution has hit geography squarely in the last thirty years, most notably in remote sensing, automated cartography and geographical information systems.

PLATE 24   Morphology of landscape. One of the earliest photographs of the complex terracing of steep-sided valleys by the Ifugao people of the hill country of northern Luzon island in the Philippines. This photograph of the central Bannawol (Banaue) district was taken in 1903. An understanding of this intricate landscape demands several strands of knowledge from the nature of precipitation and soils through to the anthropology of the complex agricultural rituals. (Source: Harold C. Conklin, *Ethnographic Atlas of Ifugao*. New Haven: Yale University Press, 1980, plate 183, p. 37)

Just such a puzzle confronted me during a period as a World Health Organization visiting Fellow at the Centers for Disease Control (CDC) in Atlanta, Georgia. I had gone there to work with statistician Keewhan Choi on the introduction of geographical elements into models of disease spread, particularly those applied to measles.[15] But a whole range of infectious diseases was being worked on at CDC and coffee-time conversations turned to the way influenza waves moved within the United States. Deaths were closely monitored each week at CDC and advice on vaccination policy given but the spatial pattern remained an enigma. Did influenza sweep across the country each winter from some distinct port of entry? Or did it arise more or less randomly across the United States with the local onset of triggering conditions? Why were there not influenza epidemics every winter? Were there 'sleeping' cases that resumed activity in the winter? The questions were not trivial since the suggestion of a national vaccination policy involved huge funding costs, not least from any legal implications of vaccine damage.

Of course maps, county by county, could be drawn for the United States on a weekly basis but these were not entirely clear. Each season looked different, each observer 'saw' a different interpretation in the pattern. The overall impression was one of randomness and confusion with perhaps a hint – but only a hint – of spatial order.

One elementary solution which was tried is shown in figure 5.7. This takes the weekly record for pneumonia and influenza deaths for a twenty-one year period from 1961 in each of the main regions of the United States (the 'test' series) and correlates this against the national pattern for the whole country (the 'control' series). A series of correlations was conducted for each region, first with the two series in phase and then with the two series out of phase. The 'out of phase' correlations were contructed by moving the test series one week, two weeks, three weeks and so on in front of the control series. The same experiment was then conducted with the test series lagging up from one to twelve weeks behind the control series. As the upper part of the diagram shows, three types of outcome are possible. First, the 'South Atlantic' pattern where the region and the nation show a peak correlation when both are in phase; as the series is forced out of phase, so the correlation falls. Second, the 'New England' pattern where one region leads the United States with a peak correlation at one week ahead. Third, the 'Pacific' pattern where a region lags the United States with a correlation peak at one week behind.[16]

By fitting an optimal curve to the weekly series, a daily peak in the correlation curve can be estimated from the weekly data. These peak figures are placed in circles in the lower map of the United States. With the time behaviour of the influenza waves plotted in space we can now add vectors to

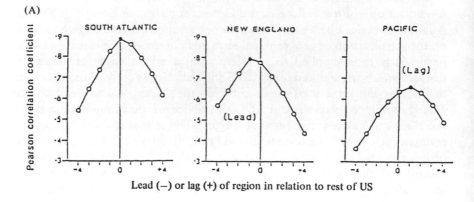

(A)

SOUTH ATLANTIC   NEW ENGLAND   PACIFIC

Lead (−) or lag (+) of region in relation to rest of US

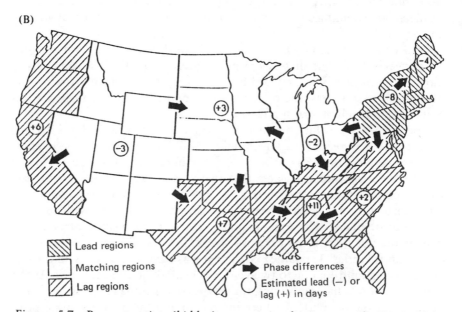

(B)

Lead regions

Matching regions

Lag regions

Phase differences

Estimated lead (−) or lag (+) in days

Figure 5.7 Reconstructing 'hidden' vectors in disease spread. (A) Lead–lag relations between three United States regions and the national weekly series for pneumonia and influenza deaths over a 21-year period from 1961. (B) Interpretation of the lead–lag relations in terms of interregional vectors to show the average flow of influenza within the conterminous United States.

Source: P. Haggett, Potential Applications of Spatial Forecasting Models to MMWR Data. Atlanta: United States Department of Health, Centers for Disease Control, CDC–CSCA Report, 1982; figs 5, 6, pp. 34, 35.

show the relative position of each region. On this basis there is a weak tendency discernible for influenza episodes to start first in the Northeast (a region which includes New York City) and then to move south and west across the rest of the country. Only the Mountain region fails to fit into this broad pattern. Such evidence is just sufficient to keep the idea of wave-like spread in contention, but not strong enough to nail the process down firmly.

Detecting the signals of these shadowy and ghostly waves from all the noise is rather like trying to tune into a weak and distant radio station. As computers allow us to build ever more powerful spatial receivers, so the geographical shape is slowly beginning to emerge. The CDC questions were later to lead to a book with Andrew Cliff and Keith Ord on *Spatial Aspects of Influenza Epidemics*.[17] Our research on the problem of epidemic spread is continuing, with help from one of the big medical charities. The most practical results so far are to suggest geographical areas where vaccination should have the highest priority in containing outbreaks.

THE GREAT DEPRESSION

The examples used have been self-indulgent insofar as they have been taken largely from research by my own group on the geography of epidemics waves. But it is important to see them as part of a much broader theme. Geographers in general are concerned with applying space–time models to a very wide range of phenomena in both human and physical geography.

Economic waves in space provide a good example of this. Among the earliest attempts to portray these was that of August Lösch, who studied the impact of the great Stock Market crash of 1929 on price waves in space.[18] By constructing indices of business activity for the local economies of a sample of counties in the American Mid West, Lösch was able to identify the time at which the main impact of the crash was felt on the local economy. By joining these times together it was possible to map the passage of the depression 'front' (to borrow a meteorological term). This showed a ripple-like shock wave moving westwards out from Chicago. For the state of Iowa, counties on the eastern border with Illinois (150 miles from Chicago) were affected in the spring months of 1930. In contrast, counties on the western border of the state (a further 300 miles distant from Chicago) did not show major falls in local activity until the late summer of 1931. This picture is shown in figure 5.8. Similar studies of the ways in which local economies have reacted to cyclical shocks, both environmental and economic, have been reported by geographer Peter Smailes in South Australia.[19] His concern has been with the way in which the delicate and fragile central-place structure of outback areas responds to two shocks; the

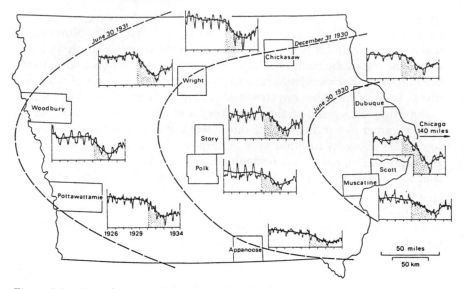

Figure 5.8   Spatial spread of an economic depression. Lösch's ideas on the spread of
the business depression of 1929–31 through the state of Iowa, central United States.
Graphs show the indices of business activity and their running means for ten
counties in the state. Stipple indicates period of consistent fall in the running mean.
Time contours show time of arrival of the depression front at various places in the
state. Note the ripple-like spread from Chicago, the nearest focus of disturbance.
Source: Lösch's general sketch reconstructed by Andrew Cliff from Iowa State Planning Board
data for 1935. See A. D. Cliff, P. Haggett, J. K. Ord and G. R. Versey, *Spatial Diffusion:
An Historical Geography of Epidemics in an Island Community.* Cambridge:
Cambridge University Press, 1981, fig. 2.6, p. 13.

drought years when wheat production is halved, and the highs and lows in
world wheat and wool prices.

BELLWETHER REGIONS?

The business cycles of sets of small areas are related to each other in a
complex way. Some areas tend to move immediately and in ways strongly
influenced by national cycles; others plough sedately through a sequence of
peaks and troughs in a different manner from most of the others. Geographers have led the search for 'early' or 'leading' areas, in the hope that
their behaviour might provide some early warning of recessions ahead.[20]

So far there is little concrete to report. Investigations of the cyclical
pattern of a number of American cities with different industrial structures

during the inter-war years found wide differences in the turning points shown by the various cities during minor economic cycles. But these differences were inconsistent in terms of leads or lags from one minor cycle to another.[21] However, during major cycles, turning points tend to be coincident.

Again, some tantalizing leads have shown up. There is weak evidence that cities concentrating in heavy industries (e.g. Cleveland and Detroit with their steel and automobile industries) tend to lead other cities for most of the period investigated. A comparison between the rest of the United States and California showed that economic activity recovered more rapidly and more fully in the latter area, suggesting that high-growth areas may show significant leads in the recovery phase of a cycle, even though coincident in the other phases.

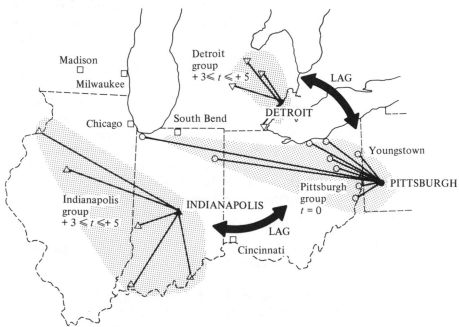

Figure 5.9 Economic impulses between cities in an integrated urban-regional system. American midwest showing strong 'within group' links with no time lags to form regional clusters and weaker 'between-group' links with time lags indicated. Eighty smaller cities with unclear group affiliations are not plotted. The five cities without bonds are linked closely to national trends rather than to the regional sub-systems.

Source: L. J. King, E. Casetti and D. Jeffrey, *Regional Studies*, Vol. 3 (1969), p. 216, redrawn in P. Haggett, A. D. Cliff and A. E. Frey, *Locational Analysis in Human Geography*, 2nd edn. London: Arnold, 1977, fig. 4.23, p. 137.

A more recent example of impact transmission modelling by geographers is shown in figure 5.9. This is based on a study of bi-monthly unemployment data for some thirty metropolitan areas in the American Mid West in the early 1960s.[22] The region chosen had traditionally high cyclical fluctuations in unemployment levels. The time period selected was one when unemployment rates were generally low, and the structural component in unemployment (that is, unemployment due to a mismatch between labour skills available in the economy and those required by industry) was relatively unimportant compared to business-cycle fluctuations.

For each pair of cities, correlations were computed between the economic series to give estimates of both the strength and the timing of inter-city linkages. Study of this correlation evidence suggested that three regional clusters of cities could be recognized. The three clusters were focused on the steel-making cities of the Pittsburgh–Youngstown area (ten cities), the agricultural machinery cities of the Indianapolis area (six cities), and the car-making cities of the Detroit area (five cities). Unemployment impulses 'hit' each group at about the same time.

Two of the clusters recognized above (those based on Indianapolis and Detroit) lagged about three to five months behind the Pittsburgh–Youngstown group, i.e. the correlations were highest when one city series lagged behind the others. Unemployment impulses 'hit' the steel-making group before the other two groups, which have different, but related, occupational structures. Some cities also had low levels of interaction with other cities in the Mid West. For example, Chicago, Madison, and Cincinnati were independent of the cyclical behaviour of their neighbours but were closely tuned in to the national pattern.

PAST REGIONAL DYNAMICS

Such economic waves are not confined to the present. Figure 5.10 illustrates work by my Bristol colleague, Leslie Hepple, on the use of a more advanced tool, spectral analysis, to disentangle the relations between regional trade-cycles in early Victorian England.[23] For this period various indicator series are available at a regional level: local agricultural prices, country banknote circulation and local bankruptcies. The bankruptcy series closely follow the overall business cycle and provide a good barometer of the health of the local economy. Details of individual bankruptcies are given in the *London Gazette* for the period and, on the basis of these series, Hepple reconstructs a detailed image of interregional linkage in the early Victorian economy.

Figure 5.10 focuses attention on the manufacturing, urban-commercial and agricultural counties of England. The Midlands counties and Yorkshire

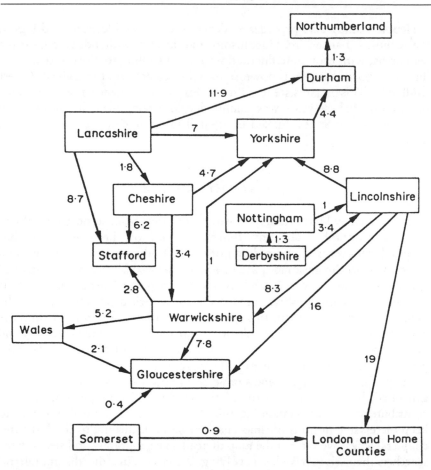

Figure 5.10 Geographical integration of regional trade cycles in early Victorian England. Phase differences between the local economies of the counties in early Victorian England. Based on spectral analysis of bankruptcy series 1838 to 1846.
Source: Leslie Hepple, 'Spectral techniques and the study of interregional economic cycles'. In Ronald Peel, Michael Chisholm and Peter Haggett (eds), *Processes in Physical and Human Geography: Bristol Essays.* London: Heinemann, 1975, fig. 19.8, p. 406.

are closely integrated but, in the north-west, Lancashire is closely linked only to Yorkshire. The north-eastern series of Northumberland and Durham is more closely linked to the Midlands group than to Lancashire. The two southern series of Somerset/Bristol and London and Home Counties are only weakly linked with the manufacturing areas. Certainly there is evidence that the regional economies are much less integrated than today, if these results are compared to regional unemployment cross-spectra for the same areas one hundred years later.

Hepple's study also provides evidence of considerable leads and lags in the economy. Lancashire, Lincolnshire and the east Midlands emerge as the lead areas, with economic fluctuations then travelling to Yorkshire and up into the north-east and down into Gloucestershire through the west Midlands. This is consistent with the historical evidence that the major depression of these years was mainly due to a textile-led slump imported from the United States, fuelled by the fall in home demand owing to bad harvests.

GEOGRAPHY AND TIME

In this fifth essay I have tried to illustrate the move from static geography to dynamic geography. The time dimension has now been added so that geographical events are unfolding not on a two-dimensional map, but in a complex multi-dimensional space in which time also plays a part. Historical geography is sometimes viewed as a separate part of geography and, as someone whose first post was in Sir Clifford Darby's department, I have the highest regard for work in this genre.[24] But, for me, the role of time has always been such an integral element – in physical, human and regional geography – that I prefer to see it as the dynamic of all our studies. All geography is, in this sense, historical geography.

I have shown through some simple cases how vectors and waves can be generated, how they move around the world, and how geographers can contribute to their understanding and, eventually, their control. We need to keep these same notions of time–space in mind in turning to the last three chapters, where we look at the map of academic geography and see how the searchers themselves have been evolving. We now adjust the 'distant mirror' and look at ourselves.

PART II
The Searchers

Geographers Through the Looking Glass

6

Family History

'Where do you come from?' said the Red Queen. 'And where are you going? Look up, speak nicely, and don't twiddle your fingers all the time.'
Lewis Carroll, *Through the Looking Glass and What Alice Found There* (1872)

Rooting around in the attic of a family home always turns up something of interest. If we think of geography itself as a multi-roomed and somewhat rambling Victorian mansion, then we too have our family bric-à-brac. Charts and globes, maps and fading aerial photographs, logs and field notebooks – now even magnetic tapes and disks – would be turned up in the search. But I would like to bring down the ladder an unusual find (see plate 15) to start this chapter. It is a circular silver plate, some ten inches in diameter, with the crest of a Cambridge college, the toothed wheel of St Catharine of Alexandria, at its centre. It contains the names of fifty geographers.

Inscribed around the periphery are the names of forty-nine geographers (all men, it being a decade later that the first women were admitted to the college) holding university posts around the world. Most come from the United Kingdom but Australia, Canada, Ceylon, Eire, Ghana, New Zealand, Uganda and the United States also figure on the list. The clue that holds the names and the locations together is inscribed at the centre of the plate. It states that it was presented to Professor J. A. Steers on his retirement on 30 September 1966 by those geographers currently holding university posts around the world whom he had taught at St Catharine's since he came there forty years before. For although this Cambridge college goes back to 1473 it was not until Steers was appointed as a teaching fellow in 1925 that its geographical history began to take formal shape.[1]

The plate is interesting for three reasons. First, it spells out the critical role of a single individual in steering the growth of a discipline; a point which could be made in equal or greater measure about Carl Sauer or Griffith Taylor. Second, it underlines the strong historical links between the leading universities of the United Kingdom, Oxford and Cambridge in particular, and the geography in many Commonwealth countries. Third, it draws a general link between individuals, groups and institutions which forms the central theme of this essay. The lifetime of any one individual on the list must be brief (and since 1966 a number have died), but the institution, already more than five hundred years old, rolls on into the next century with a different group of names ensuring its continuity.

For the arrows of space, the vectors which we looked at in the last chapter are not confined to the spread of diseases, or economic links, or inflation waves. They also apply to the spread of intellectual information in the shape of ideas. These spread not only freely from individual to individual, mainly through the printed page, but also in a more channelled flow by lecture, by tutorial, by example, by shared expeditions, by membership of a department, by correspondence and so on. In this essay I want to look at the way geographical ideas move through the complex genealogical space which is formed by the meshing in time and space of individuals, groups and institutions, and how the speed and efficiency of that movement is being changed. Geography as a discipline is like a family with all the unwritten conventions, fierce loyalties and hidden enmities that family ties imply.

CHAINS OF INDIVIDUALS

In this historical web the individual life is woven into larger structures. I have chosen to present this weaving process as though it took place in three acts of a long-running play. Act I is dominated by sporadic research activity carried out by only a few geographical scholars, and these threads are isolated in time and space. In the second Act individuals begin to weave themselves into groups and societies and the research effort is greater in volume and more continuous over time. By Act III, geographical research is bound into a tightly woven enterprise which is still greater in volume and more continuous over time. Now geographical research is incorporated into state-supported organizations (such as colleges and universities) and geographers in different countries become linked together through international institutions.[2]

Clearly the three acts cannot be precisely fixed in time. Like stage plays, the same drama may be running at different theatres around the world with different actors and with a slightly different textual interpretation. In the earliest act there is a single lifeline of short duration; in the later a thick, interwoven cable of lines in which the institution outlives the life-strands that make it up in different historical periods.

The first growth period, from the beginning of formal geographical study in ancient Greece to the mid-nineteenth century, shows a pattern of geographical studies which was sporadically distributed in time–space.[3] The number of scholars who would have described themselves as geographers was always small, and it was only occasionally that clusters of workers formed – as in the great library of Alexandria in the second century BC, in the navigational schools of Portugal in the fifteenth century, or in the atlas-making houses of the Low Countries in the sixteenth century. The patronage that enabled these groups to join together usually came from the need to solve practical geographical problems; such as methods of surveying the earth, instruments for marine navigation, map making of newly explored areas, and the printing of world atlases.

In this early and fragmented period, geographers found the answer to many questions about the shape of the earth and ways of putting spatial information on maps. However, most geographical schools were shortlived, and had fluctuating fortunes. By the close of the eighteenth century, the general rise in scientific activity in the academies of western Europe allowed more individual strands to be recognized. The chart in figure 6.1 is an attempt I made some years ago to plot the lifelines of many of the leading scholars in geography in the two centuries from 1775. Remember, in interpreting this diagram, that the number of names that might have been included was far less before 1800 than it was after 1900: the diagram is more selective as we move from left to right. Indeed, as we shall see later, the growth of geography over the two centuries spanned by this diagram has been exponential. Most of the geographers who ever lived are alive today! Note too that in the earlier period boundaries between disciplines were loosely drawn. It is not surprising that individuals like Immanuel Kant, Alexander von Humboldt and T. R. Malthus played notable roles in the growth of several other fields: philosophy, biology and demography, as well as geography. The diagram also underlines the significant part played by Germany in all the early part of the period. (Until 1939 nearly half of all geographic publications were in the German language.) Inevitably too, the diagram reflects the social and spatial biases of the period. There are very few women geographers in the list, and certain countries isolated by language from the Western mainstream (notably China) are under-represented.

The exact significance of any one individual in the overall growth process is difficult to assess. Science often involves a snowball effect, because of which more than due emphasis is placed on the contributions of a few individuals; for example, on the work of the few physicists or chemists who have won the Nobel Prize. This leads to the so-called 'Matthew effect' (after the Gospel's remark that 'to him that hath more shall be given').[4] This also occurs in geography, and figure 6.1 inevitably reflects it. Yet it would be impossible to think of German geography in the mid-nineteenth century

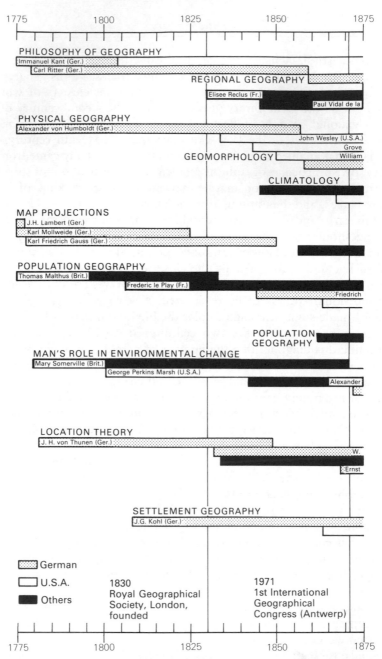

Figure 6.1   Geography, 1775–1975. The chart shows some leading scholars in the most recent period of geographic research. It indicates the emergence of some major schools and the changing national balance of research. (Compare the German names in the nineteenth century with the American ones in the twentieth.) The geographers' names have been mainly taken from a list in Sir Dudley Stamp's Longman's *Dictionary of Geography* (Longmans: London, 1966), largely concerned with western scholarship. No comparative information was available on the growth of, say, Chinese geography. Some names have been included to illustrate the major contributions to geographic thought by scholars from other disciplines, particularly

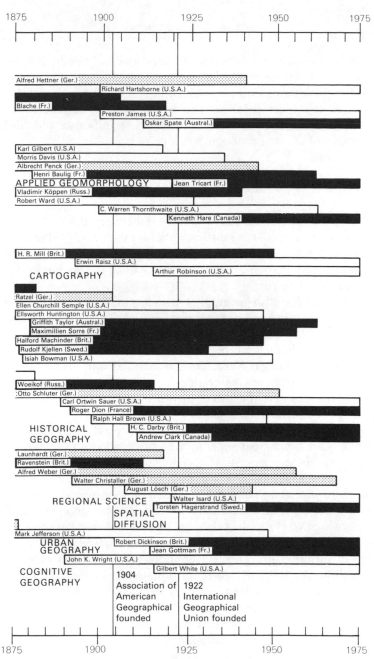

	1875	1900	1925	1950	1975

Alfred Hettner (Ger.)
Richard Hartshorne (U.S.A.)
Blache (Fr.)
Preston James (U.S.A.)
Oskar Spate (Austral.)

Karl Gilbert (U.S.A)
Morris Davis (U.S.A.)
Albrecht Penck (Ger.)
Henri Baulig (Fr.)
APPLIED GEOMORPHOLOGY   Jean Tricart (Fr.)
Vladimir Köppen (Russ.)
Robert Ward (U.S.A.)
C. Warren Thornthwaite (U.S.A.)
Kenneth Hare (Canada)

H. R. Mill (Brit.)
Erwin Raisz (U.S.A.)
CARTOGRAPHY   Arthur Robinson (U.S.A.)

Ratzel (Ger.)
Ellen Churchill Semple (U.S.A.)
Ellsworth Huntington (U.S.A.)
Griffith Taylor (Austral.)
Maximillien Sorre (Fr.)
Halford Machinder (Brit.)
Rudolf Kjellen (Swed.)
Isiah Bowman (U.S.A.)

Woeikof (Russ.)
Otto Schluter (Ger.)
Carl Ortwin Sauer (U.S.A.)
Roger Dion (France)
Ralph Hall Brown (U.S.A.)
HISTORICAL   H. C. Darby (Brit.)
GEOGRAPHY   Andrew Clark (Canada)

Launhardt (Ger.)
Ravenstein (Brit.)
Alfred Weber (Ger.)
Walter Christaller (Ger.)
August Lösch (Ger.)
REGIONAL SCIENCE   Walter Isard (U.S.A.)
SPATIAL   Torsten Hagerstrand (Swed.)
DIFFUSION
Mark Jefferson (U.S.A.)
URBAN   Robert Dickinson (Brit.)
GEOGRAPHY   Jean Gottman (Fr.)
John K. Wright (U.S.A.)
COGNITIVE   Gilbert White (U.S.A.)
GEOGRAPHY

1904 Association of American Geographical founded

1922 International Geographical Union founded

earlier in the period when the boundaries of fields were loosely drawn. Names have been chosen with a view to illustrating the emergence of significant research themes. Inevitably, other geographers would choose other names. Yet, while there might be only a modest overlap, the same general pattern of evolution of research themes over time would probably emerge. Note that the country given after each geographer's name is sometimes nominal since he or she may have worked for long periods in two or more countries.

Source: P. Haggett, *Geography: A Modern Synthesis*, 3rd revised edn. New York: Harper and Row, 1983, fig. 25.2, pp. 606–7.

without taking into account the work of Carl Ritter and Friedrich Ratzel; or France without Vidal de la Blache; or the United States without W. M. Davis; or Britain without Halford J. Mackinder. On the present century our perspective is too short to allow the key figures to be discerned with certainty. In any case, team research is growing in significance and this blurs the importance of the single scholar working in isolation.

A LIFE IN GEOGRAPHY

I have drawn an analogy earlier in this essay between the life of any one geographer as a thread running through a fabric with many other lives and events woven across it. We can illustrate this most easily by a specific example and I have chosen a name we met earlier in this book: Torsten Hägerstrand (see plate 16).

Hägerstrand is a Swedish geographer who has made major contributions to two research areas in human geography: spatial diffusion and time–space geography.[5] Hägerstrand studied at Lund University under Helge Nelson and was strongly influenced by the mathematical approaches of the Estonian geographer, Edgar Kant. The first scholar introduced him to the meticulous tracking of human migration movements in small areas of rural Sweden, and the second to the idea of how probability theory could be used for the study of settlement patterns.

The two strands came together in his 1953 doctoral thesis *Innovationsförloppet ur Korologisk Synpunkt*, in which the adoption of agricultural innovations by farmers in central Sweden was conceived as a series of diffusion waves whose passage could be mapped, modelled and simulated. The ideas there were elaborated and extended in *Migration in Sweden*. This early work pioneered new methods in cartography and Monte Carlo simulation, which were rapidly taken up by North American colleagues in the late 1950s and widely adapted and extended, particularly by computer modelling of diffusion processes. Over more recent decades Hägerstrand has moved from the study of aggregate time–space studies to the detailed dissection of individuals' movements over very short time periods. Hägerstrand and his Lund colleagues have been able to show how an understanding of changes in our trajectories at a microscale allows a much clearer understanding of the changes in the aggregate pattern of human population distribution.

Why did Hägerstrand choose these topics rather than the hundreds of others within geography? Clearly, formal university courses and other scholars are strong influences. But childhood and the environment in which we were raised seem to have played some part as well. Son of a rural schoolteacher, Hägerstrand has written at length on his own childhood and the natural and social history of the rural Swedish environment of his home village:

> The destinations of family walks and solitary excursions were either hilltops with a wide prospect over the valley or some exceptionally large trees which had their own names and were regarded somewhat as silent friends. Some very big erratic boulders played a similar role. The whole of nature was not just nature as an undifferentiated environment. It was a room inhabited by personalities. For my memories of this world, images of touch and smell and sound come side by side with the visual pictures. I can still feel in my body how it was to sit on a certain stone or climb in this or that tree.

And, again, on social events in the village:

> Every night at eight o'clock, the year round, a crowd of people assembled at the railway station while the north-going train made a short stop. The most common excuse mentioned was the purchase of an evening newspaper or journal. But in fact, the gathering had the same meaning as the 'ramblas' in South-European cities – a time of belonging ... When the sound of the train faded away behind the endless screens of spruce, we had once again been reconfirmed that we were anchored in an ordered, wider world.[6]

Later influences in his life are captured graphically in a diagram that accompanies an autobiographical essay.[7] This is reproduced in figure 6.2 and shows at the far right the stages in Hägerstrand's own life (from a small boy in central Sweden to a research professor at Lund) and on the far left the major historical events in Europe between 1910 and 1980. Between these two markers the diagram plots the different influences on his work, the key books he read, the places he visited, the colleagues he worked with, and the 'gate openers', those who opened up opportunities for him to move forward in research and teaching.

I have taken Torsten Hägerstrand as an example since he is the geographer who particularly influenced my own research. But, increasingly, biographical studies of geographers are now being published and there is a splendid range of material appearing which allows the emergence of scientific ideas to be placed in a firm historical context.[8]

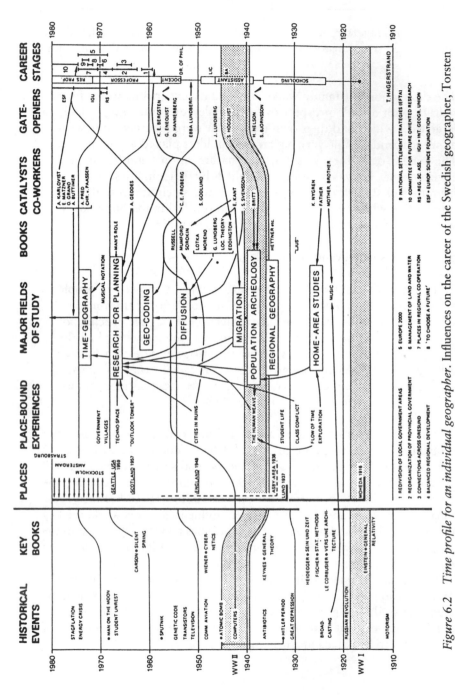

Figure 6.2 Time profile for an individual geographer. Influences on the career of the Swedish geographer, Torsten Hägerstrand.

Source: Anne Buttimer (ed.), *The Practice of Geography.* London: Longman, 1983, fig. 16.3, p. 255.

GEOGRAPHICAL SOCIETIES

If we now return to my three-act view of our history, then a second period of growth is marked by the spatial interlinking of research. This begins at the start of the nineteenth century in the most developed countries when the growth of postal services and improved communication brought about much faster interchange of notes and correspondence between scholars. Improved printing methods reduced the costs of books, journals, atlases and maps and encouraged the build-up of private libraries.

One of the earliest methods of linkage was forming small groups and societies to foster the common interests of a few like-minded individuals in geography. Often these groups were transitory, meeting in coffee house or club to share tales of travel and exploration. In London the Travellers' Club was founded in 1819. The qualification for election to the Travellers' was not Arctic exploits, Saharan crossings or Himalayan climbs but, more modestly, 'to have travelled out of the British Isles to a distance of at least 500 miles from London in a straight line'. Such a demarcation conveniently allowed wintering on the French Riviera or a summer villa in Tuscany to count as 'travel'. As one wit commented:

The Travellers are in Pall Mall, and smoke cigars so cosily,
And dream they climb the highest Alp, or rove the plains of Moselai.
The world for them has nothing new; they have explored all parts of
   it;
And now they are club-footed (!) and they sit and look at charts of it.[9]

But a genuine interest in overseas geography was fostered and many members of more academic societies sought its membership. It still stands in Pall Mall and remains a favourite haunt of a few geographers. Still older is the Geographical Club which still meets regularly and is a major focus for British geographers.

Such informal groups were the forerunners of the national societies that emerged in the early years of the nineteenth century. These societies usually had a strong interest in global exploration. For instance, the Royal Geographical Society in London dates from 1830 and marks the merger of several early exploring clubs like the Association for Promoting the Discovery of the Interior Parts of Africa, founded in 1788 (plate 17).[10] Likewise the American Geographical Society of New York was founded in 1852 by a group of businessmen to provide a centre for accurate information on every part of the globe.[11]

As geography began to grow in colleges, so a group of societies emerged.

These were the national professional groups, largely dominated by university and research geographers. These societies are later in date of formation, smaller in membership and less catholic in scope than the national societies. The Association of American Geographers (founded in 1905), the Institute of British Geographers (1933) and the Institute of Australian Geographers (1960) are typical of this group.[12]

A third group of societies was founded primarily to promote geographic education in schools: the Geographical Association in Britain and the National Council for Geographic Education in the United States are examples.

Last come the small and specialized geographical sub-groups. These organizations are the most recent and the smallest, but also currently the most vigorous of the four types. Many originated during the 1950s as sub-groups within national professional organizations concerned with a particular aspect of geography (e.g. cartography, geomorphology or quantitative methods). This trend for some societies to split up once a certain size has been reached is illustrated by the formation of an increasing number of specialist groups by the Institute of British Geographers.

THE NATIONAL GEOGRAPHIC

Although most countries followed the pattern described, one national society stands apart from the rest. The National Geographic Society was founded in 1888 in Washington DC to 'increase and diffuse geographic knowledge and to promote research and exploration'.[13] It is now one of the world's largest non-profit, scientific and educational institutions. Today its members total eleven million and it is estimated that its leading monthly journal, *The National Geographic*, is read by approaching forty million readers. Almost from its foundation the Society sponsored explorations and research projects in the remoter areas of the earth. Its first expedition explored and mapped Mount St Elias along the Alaska–Canada border. It led the conservation movement long before the cause was a popular one, and as early as 1915 the Society and its members contributed $100,000 to help preserve 2,000 acres of Sequoia National Park when the giant Californian sequoias were threatened by lumbering.

The Society has led the way in taking the mission of geographic understanding out to the public. It began a specialist map service in 1918, and produces its own atlas and globes. For the last quarter-century it has produced special films for television and is increasingly involved in the educational field with the production of multi-media material for schools. In 1975 it launched its second journal, the *National Geographic World*, aimed at children in the eight to twelve year age group, and this now has a circulation of around two million. Recently a third journal, *National Geographic Research*, has been added.

The size and popular success of the National Geographic Society put it on a scale quite unlike that of any other geographical society or indeed of other scientific societies. It has achieved this by building on the innate curiosity of most people about their world, by adopting a wide and pragmatic view of geography, and by pioneering very high quality of printing and publishing. In retaining a clear view of its public and its mission to that public, the National Geographic provides a fascinating counterweight to the more scholarly but, to the public, more obscure aims of the other geographical societies.

Taken together, geographical societies of all four types have been growing very rapidly. One estimate suggests that their numbers have been doubling every forty years since the late nineteenth century.[14] Numbers of such societies worldwide are difficult to estimate but they run into many hundreds. They range from well-established societies with a lifespan of more than 150 years and millions of members, to small ephemeral groups of student members which wax and wane. Plate 18 gives an example of the latter, showing the total membership of the Vidal de la Blache Society as it stood in the summer of 1984.

THE JOURNAL SUPERMARKET

The prime function of the geographical societies was to foster common research interests through the reading of papers and the publication of their transactions in the form of regular journals. The establishment of journals like the *Annals* of the Association of American Geographers in 1910 represented key breakthroughs in the circulation of research findings. Other journals were published by interested individuals, as was Petermann's *Geographische Mitteilungen* in 1855. Still others were founded by small groups of research workers in a single department, as was Ohio State University's *Geographical Analysis* in 1969.

The sheer growth in numbers of these journals provides a rough but useful index of the increasing volume of geographic research. As figure 6.3 shows, the field has been expanding regularly ever since the seventeenth century. This shows that the number of all scientific periodicals doubled about every fifteen years. The number of geographical periodicals also doubled regularly, but at about half that rate.[15] The slower rate of increase in geography is typical of older, established sciences (like geology, botany or astronomy), because the total increase in all scientific publications reflects

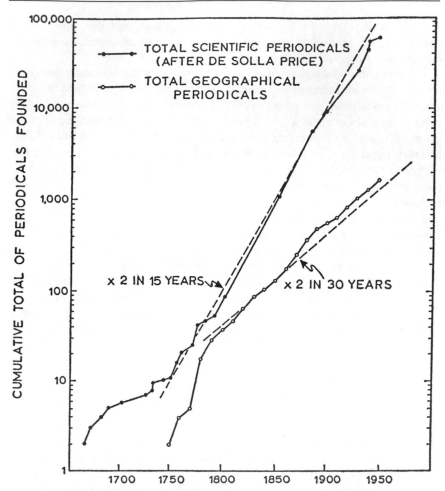

Figure 6.3 Growth of geographical journals. The chart shows the cumulative total of scientific and geographic periodicals founded since the middle of the seventeenth century. The vertical scale of the graph is logarithmic: the number of both scientific and geographical periodicals has been growing exponentially.

Source: D. R. Stoddart, 'Growth and structure of geography'. *Institute of British Geographers, Transactions,* Vol. 41 (1967), pp. 1–20.

also the birthrate of wholly new scientific fields like molecular genetics, computer science or AIDS research.

For geography in Britain, there is a strong contrast between the large number of new serials and the relatively few journals published before 1960. Typical of the scores of new serials of the last thirty years are those with broad discipline-wide coverage like: *Geoabstracts, Area* and *Progress in Geography* (now two separate journals), and specialized journals within geography such as the *Journal of Historical Geography, Journal of Biogeography* and *Teaching Geography*. These are supplemented by inter-disciplinary journals with a strong geographical flavour: *Urban Studies, Regional Studies, Environment and Planning, Journal of Environmental Management* and *Earth Surface Processes* – as well as numerous limited-circulation publications from departments.[16]

The importance of the different journals has not remained constant. New journals, like new products in a supermarket, may sometimes steal part of the market share of an existing journal. Equally a journal may have changed its content and readership to retain an ecological niche within the scholarly marketplace. One of my old students, Tony Gatrell, has captured this shifting role of journals by plotting them in the multi-dimensional space

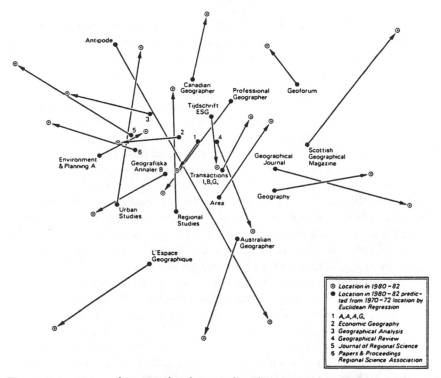

Figure 6.4   Maps of geographical journals. Changes in the relative position of geographical journals in multi-dimensional space.

Source: A. C. Gatrell and A. Smith, 'Networks of relations among a set of geographical journals'. *The Professional Geographer*, Vol. 36 (1984), pp. 300–7.

using the techniques which we met in chapter 3 of this book.[17] Figure 6.4
gives an example of his findings.

Most difficult to track are the occasional publications and discussion
papers of the more active research departments. These are produced in small
numbers, often with a strongly targeted distribution to other workers. They
provide a useful means of keeping together those loose networks of
scholars, the 'invisible colleges' of researchers who need to keep each other
abreast of the most rapid developments in the field.[18] With the advent of
desktop publishing packages for microcomputers, electronic mail, fax
facilities and interlinking computer networks, the future for geographical
publishing has rarely been more difficult to foresee. The pattern and speed
of transfer of research may well shift significantly by the year 2000.

GENETIC MAPS OF IDEAS

I illustrated in the last chapter some of the ways in which geographers have
tried to understand the way infectious diseases are transmitted in space. But
many of the models developed there are part of a more general family which
apply to a wide range of spread phenomena. Ideas are also infectious, and
the spatial transmission of information has also attracted the interest of
diffusion modellers.[19] If geographers are concerned to see their ideas widely
debated and, if justified, accepted, then it is increasingly important to ask
which journals are likely to prove the most effective as diffusion channels.

We can envisage each paper as a pebble tossed into a fast-running river
and causing splash waves. These may be swept downstream for a few
moments, or cancelled immediately by a shower of other pebbles, as new
papers are thrown in. One of the ways of measuring this river of
information flow is to look at citations. Each scientific paper carries a set of
references on the final page (just as this book has notes at the back). The
traditional rules of scholarly writing mean that all authors are under an
obligation to say from where we have drawn our ideas.

So the trail of references or citations to previous literature forms a
network linking one paper to another in a sort of genetic map across which
ideas, rather than individuals, are conceived and passed on. Links from each
paper will be backward and forward: backwards to the older papers cited,
forward as the paper is cited by other writers in their own writing. So the
printed page takes on a life independent of the author and, in the case of the
classics, may be cited for decades after their authors have died.

Unravelling the network of cross-references is the task of citation studies,
a branch of science-watching which is as intriguing as it is controversial.[20]
The largest study of citations has been carried out in the United States by the

Institute for Scientific Information (ISI) in Philadelphia. This publishes three annual sets of volumes, the Science Citation Index, the Social Science Citation Index and the Arts and Humanities Citation Index. Some idea of the size of the enterprise is given by the fact that it now runs to over twenty volumes which cover 8,000 journals. Since each journal has an annual average of around 120 papers, and since each paper carries an average of between twelve and thirteen references, the total number of citations tracked exceeds ten million for each recent year. Geography is mainly included in the social science citation volumes with nearly thirty specifically geographical journals now covered (e.g. *Geographical Review*) and a larger number of related journals in which geographers dominate the contributions (e.g. *Regional Studies*). Still other geographical journals are analysed in the science citation volumes.

One critical and much-debated concept in the ISI citation programme is the 'impact factor' of a journal. For any journal this relates to the number of times its papers are cited by other journals (the *export* of its ideas) divided by the number of papers it cites as footnotes to its own papers (the *import* of ideas from other journals).[21] A journal with a ratio of exactly one would have a precise balance between imports and exports.

For a recent year I looked at these ratios and found that most English-language journals in geography are characteristically importers of ideas from other fields with ratios between one half and one. We need to recall that this is for a single year and that, of course, geographers publish widely in journals which are not specifically geographical in name. Here environmental journals such as the *Journal of Atmospheric Science*, *Ecology* and *Water Resources* characteristically have impact factors around two, higher than those on the social science side. Cartographic journals appear to fall between these two groups.

RALLYING AROUND THE FLAG?

So choice of journal faces us with a dilemma. By strongly supporting the leading geographical journals and ensuring that their most important findings are first printed in these journals, geographers keep ideas within the family and increase the professional identity of the field. But, like scientists in other fields, we need to report to a wider audience. To achieve the maximum circulation for geographical ideas, geographers need to publish more frequently in the general scientific journals which have very wide circulations and high impact factors. Two such journals are the weeklies *Nature* and *Science* which have a sixfold balance of exports over imports. The popular journal *Scientific American* also has a threefold balance. My own experience with a paper published in one of these journals[22] suggests that the offprint requests which are generated are much greater than from specialized journals.

So far we have assumed that the main transmission of geographical ideas is through journals. One of my Bristol colleagues, Professor Neil Wrigley, has investigated the role of books as well as journals within geography. He finds that, unlike many of the sciences but like parallel fields in the humanities, the seminal book has played a critical role in shaping how geographers see their field and how they pursue their research problems.[23]

UNIVERSITY DEPARTMENTS

A critical role of geographical societies was to convey the importance of the problems their members were studying to the rest of the community. Their partial success was marked by the beginning of a third phase of geographical study, overlapping the second, in which geography departments were formally established in major universities (plate 19). In Britain, the leading society – the Royal Geographical Society – played a key role in lobbying for the subject and in establishing geography at both Oxford and Cambridge.[24]

In my view, the university department forms in the twentieth century a uniquely effective institution for furthering geographical study. I see its role as fourfold. First, it is a 'frontier post' from which the advancement of new geographical knowledge (through research) can be pushed forward. Second, it is a 'school' for transmitting that knowledge through teaching, not only to undergraduate and graduate students but through extension lectures and through publications. Third, it is a 'monastery' (at least in the medieval sense) where geographical knowledge is conserved (through libraries and archives) and where a learning environment is created in which staff and students can deepen their geographic interests. Fourth, it serves as a 'district hospital', a resource centre with skills, knowledge and technical resources which can be called upon by the community, whether local or global, for advice and consultation. The mix of the four roles will vary from one university to another and from country to country.

In Europe, Germany took the lead in establishing geography as a university subject, with an important watershed in 1874. With Bismarck's encouragement, the Minister for Education in Prussia issued a directive that each university in the state should establish a chair in geography.[25] Up to that time only five universities in Germany (Berlin, Brelau, Gottingen, Halle and Leipzig) had professorships of geography. By the end of the century, nineteen of the twenty-one universities in the German empire offered instruction in geography.

Developments in France were only slightly less rapid, but the United States, Britain and the Commonwealth lagged considerably. New geography departments often showed an irregular spatial diffusion pattern, marked by curious regional concentrations and sparse areas; figure 6.5 shows the distribution of degree-giving departments in the United States, where there is presently a strong midwestern emphasis.[26]

Britain offered facilities for the study of geography at only two of its universities, London and Oxford, prior to World War I. Geography departments were added to that list in two surges, following each of the two world wars. My own university, Bristol, typifies the first of those waves.[27] Medical teaching in Bristol goes back to the eighteenth century but a formal University College was not established until 1876. The first holder of the chair of geology took a keen interest in geography and a first lectureship in that subject was established within the geology department in 1920. Typically, the holder of the post was a graduate not in geography but in one of two related fields, geology and mathematics. The establishment of a department (1925), of chairs of geography (1933) and the build-up of staff are shown in figure 6.6. Today it has eighteen academic staff, a dozen research staff, ten technical, secretarial and library staff, twenty to thirty graduate students and about 170 undergraduates.

The Bristol pattern of linkage with an earth science was a common one in many countries. The United States presents a number of examples. Although courses in geography were offered at the University of Wisconsin, Madison, from 1862, they were developed and taught as a continuous programme by Rollin D. Salisbury from 1891. Instruction was first given in the Department of Geology, Mining and Metallurgy, and Geography did not emerge as a separate department until 1928.[28] Geography in the University of California at Berkeley is also one of the older departments, dating from 1898, but it shared its base with geology until 1923.[29]

Both Wisconsin and California illustrate the role of the single, long-serving scholar in shaping and integrating a university teaching and research programme. Glen Trewartha and Carl Sauer influenced their respective departments for decade after decade: Trewartha on the Madison faculty from 1920 to 1967, Sauer on the Berkeley faculty from 1923 to 1957.

INTELLIGENCE IN WAR AND PEACE

The role of geographers in providing regional intelligence for the state extends from Henry the Navigator to the CIA. During time of war, the need for geographical information becomes still more acute. Groups of geographers were brought together to serve in areas from photointelligence to

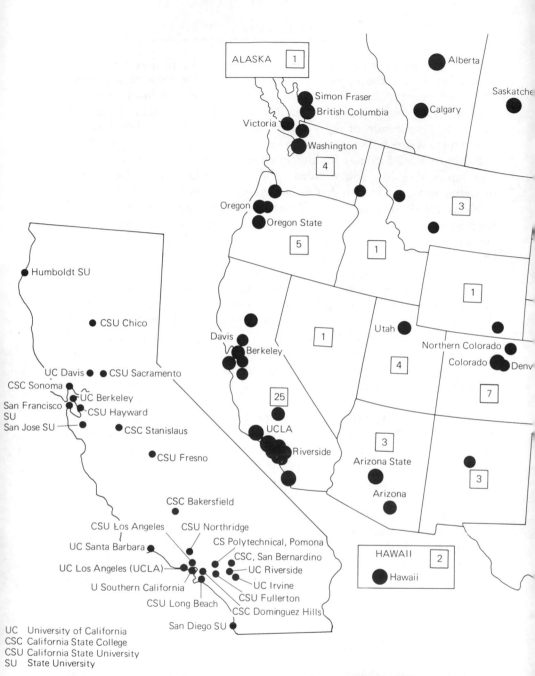

Figure 6.5 Spatial distribution of geography departments. A cross section of geography in North American universities in the mid 1970s. (Left) Geography departments offering an undergraduate major in the state of California. (Right) Location of geography departments in Canada and the United States offering courses at the graduate level. Those with doctoral programmes are named. Where the number of faculty was not available, the location is shown by the smallest of the four circles. Maps are based on data given in *Guide to Graduate Departments of*

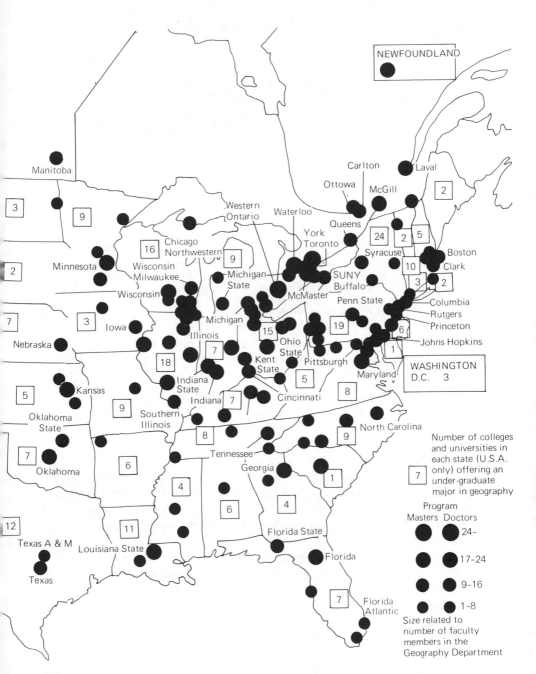

NEWFOUNDLAND

Manitoba

3

9

16 Chicago
Northwestern

Minnesota

2

Wisconsin
Milwaukee

Wisconsin

7

3

Iowa

Nebraska

18

5 Kansas

Oklahoma
State

9

7

Oklahoma

6

12

Texas A & M

11 Louisiana State

Texas

Western
Ontario

Waterloo

York
Toronto

9 Michigan
State

Carlton

Ottowa

McGill

Queens

24

Laval

2

2 5

Syracuse

Boston
Clark

10

3 2

McMaster

SUNY
Buffalo

Penn State

Columbia
Rutgers
Princeton

15

Michigan

Illinois

Indiana
State

7

Indiana 7

Ohio
State

19

6

Johns Hopkins

1

Kent
State

Pittsburgh

WASHINGTON
D.C.   3

5

Cincinnati

8

Southern
Illinois

8

Maryland

North Carolina

Tennessee

9

Georgia

1

4

6

4

Florida State

Florida

Florida
Atlantic

7

Number of colleges
and universities in
each state (U.S.A.
only) offering an
under-graduate
major in geography

7

Program
Masters Doctors

24-

17-24

9-16

1-8

Size related to
number of faculty
members in the
Geography Department

Geography in the United States and Canada 1976–1977, (Association of American Geographers, Washington, DC, 1976). Some universities not shown on this map now offer geography programmes and vice versa. The Canadian coverage is less complete than that for the United States. [Map from drawing by Andrew Haggett.] Source: Peter Haggett, *Geography: A Modern Synthesis*, 3rd revised edn. New York: Harper and Row, 1983, fig. 25–6, pp. 612–13.

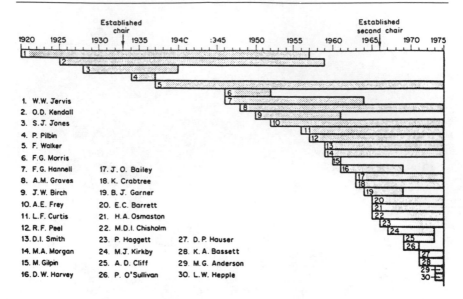

Figure 6.6   Lifelines of members of a geography department. The academic staff within a typical geography department, the Department of Geography in the University of Bristol from the first staff appointment in 1919 up to 1975.
Source: Ronald Peel, 'The Department of Geography, University of Bristol, 1925–75'. In Ronald Peel, Michael Chisholm and Peter Haggett (eds), *Processes in Physical and Human Geography: Bristol Essays.* London: Heinemann, 1975, fig. 20.1, p. 415.

military mapping. Part of the work in World War II remains classified but we have some scholarly legacy in the volumes of the Admiralty Handbooks produced by the UK's Naval Intelligence Division. Similar regional handbooks were produced by geographers working with intelligence agencies in the United States.[30]

Outside the short-term demands of war, the need for national geographical information has led a few countries to the emergence of separate institutes outside the university sector. For example, in Brazil the Instituto Brasileiro de Geografía e Estatística (IBGE) and in the Soviet Union Nauk SSSR are each charged with the investigation and publication of regional data within their vast national territories. The latter has a staff of over 300 geographers working in ten divisions and a massive publishing programme, including the bi-monthly *Izvestiya, Seriya Geograficheskaya*. Even when separate geographical institutes have not been set up, geographical research has been increasingly incorporated into such organizations as Britain's Ministry of Town and Country Planning (now part of the Department of the Environment) and Australia's Commonwealth Scientific and Industrial and Research Organisation (CSIRO).

INTERNATIONAL NETWORKS

Since 1922 the initiation and co-ordination of geographical research requiring international co-operation have been handled through the International Geographical Union (IGU). Like the Olympic Games, this organization holds international congresses at intervals of four years. The most recent was in Australia; the 1992 meeting will be in the United States. Between congresses it appoints specialist commissions to study subjects like arid zones, quantitative methods or economic regionalization (plate 20). If we compare member countries, we can see that the size of their geographical research effort is broadly related to their overall scientific budget. However, smaller countries play a role in research out of proportion to their size. One outstanding example is Sweden which, with few universities and a small group of geographers, has led research in several important areas. The volume and quality of research from New Zealand are also remarkably high in relation to its small number of research centres.

We have seen in this chapter how the scholar, the society, the journal, the national institution and the international body form a sequence in the family history of geography. As one moves further along the sequence so the size of the group and its longevity increase. In an ideal world, larger institutions form the matrix within which individual scholarship can flourish; this was the ideal of the 'community of scholars' embodied in an Oxbridge college. But size and stability are bought at a price. Corporate structures tend to make corporate policy and balancing this with the different needs posed by individual researchers is one of the themes that runs through the next two chapters.

Looking back at the roots of geography shows the importance of family ties: we all inherit more than we realize from our predecessors. But nostalgia is best in small doses. As the Red Queen says, 'And where are you going?' New horizons beckon, and for every generation of geographers a distinctive view of the world has always to be won anew.

7

Shifting Styles

Still the Queen kept crying 'Faster!' but Alice felt she could not go faster, though she had not the breath to say so. The most curious part of the thing was, that the trees and other things around them never changed their places at all: however fast they went, they never seemed to pass anything. 'I wonder if all the things move with us?' thought Alice.
Lewis Carroll, *Through the Looking Glass and What Alice Found There* (1872)

As Mrs Thornburgh is always reminding us, it's 'propinquity does it'.[1] Until the late fifties, the geographers at Cambridge shared their lovely old building (a former School of Forestry and timbered accordingly) with the Department of Geophysics. This meant that visitors to Sir Edward Bullard's distinguished group often gave general lectures to which interested geographers could also come. So it was that growing up in geography there gave a unique opportunity to attend lectures from the world's leading earth scientists, oceanographers like Maurice Ewing from Woods Hole or Francis Shepherd from the Scripps Institute of Oceanography. Still wider lectures on the physics of the earth were given in the university at that time by Sir Harold Jeffreys.[2]

What is remarkable in retrospect is that, although Cambridge geophysicists were to play a leading part in the plate tectonics revolution of the 1960s, I can recall little hint of support for continental drift in those 1950s lectures.[3] Alfred Wegener's ideas that the continental blocks were mobile and that the shield areas of South America, Africa, India and Australia had once formed part of the gigantic super-continent of Gondwanaland had been current since 1912. But they were regarded for the next half-century as fanciful, even physically impossible.

Yet the switch from a stabilist view to a mobilist view of the earth's crust was to occur within a year or so. Even more oddly, much of the 'new' evidence for the reinterpretation had been available for twenty years. As we saw in the first chapter with the *Gestalt* diagrams in psychology, it is the

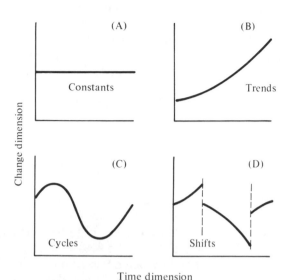

Figure 7.1 Types of change in geography. (A) Constants (B) Trends (C) Cycles and recurrences (D) Shifts or discontinuities. See discussion in text.

switch to seeing a new pattern which is the trigger to sudden change.

Such switches recall a cartoon in the *New Yorker* showing a father putting the finishing touches to the name of an upturned boat (see plate 21). His paintbrush hovers over the final 'h' in the title – 'Sea Witch' – and he is obviously pleased with a careful job nearly completed; each letter exactly matches the style of the others. Unfortunately, the boat is upside down but only when it is put into the water will the gross error be clear. An unhappy small boy is watching the process and is given the line of the cartoon caption: 'Dad, do you want to know something?'

The relation between one generation of geographers and another reminds me of that cartoon. So often we work on a succession of research projects that clearly are carefully articulated with one another, like the letters on the side of the boat. Since the internal logic and precedent are well established, there is little questioning of the whole enterprise of which the individual projects form a part. The fundamental questions are asked only rarely, and then usually by someone from a younger generation.

I see those sudden changes or 'flips' in the direction of a discipline as only one of a series of changes that characterize growth (see figure 7.1). Two others are the long-term monotonic changes referred to in this essay as trends and the recurrent or pulsing changes (described as cycles). Trends, cycles and shifts – all changes of different abruptness and complexity – form the triple structure of this chapter, where we look at each in turn.[4]

MEASURING CHANGE IN A DISCIPLINE

Estimating changes in any scientific field poses severe measurement problems. It would be splendid if we could simply count up the geographical knowledge at two periods and note the changes. But if we assign a particular piece of knowledge a value, it is not additive since its combined value when united with something else will depend very much on what that something else is.[5] Also knowledge is irreversible; it cannot logically become unknown once it is known, though it may come to be forgotten over a long period. It is subject to infinite multiplicability: if knowledge is gained by one geographer, it is not diminished by being passed on to ten others. Clearly, counting the number of geographical ideas is not likely to work.

Another approach is to use 'proxy' measures. Drawing on the methods pioneered by the American historian of science, Derek de Solla Price, David

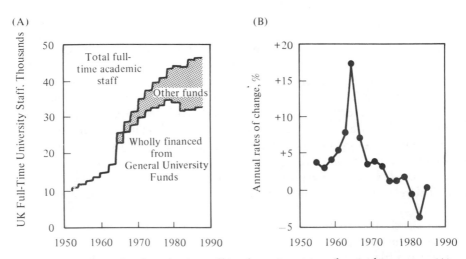

Figure 7.2   Growth of academic staff in the universities of a single country. (A) Changes in academic staff in United Kingdom universities. The shaded area represents academic staff supported on research grants and contracts. (B) Biennial changes in the number of tenured academic appointments. The rapid growth in the mid 1960s is clearly shown.

Source: Graphs drawn from annual returns published by the University Grants Committee.

Stoddart showed that one of the simplest of such proxies to use is the number of scientific journals published in a field.[6] As was described in the previous essay, for all scientific journals the rate of growth over the last three centuries is equivalent to a doubling every fifteen years and for geography (and other older sciences) half that rate. A second measure of magnitude is the number of specialist geographical societies: geographical societies have doubled every forty years since the late nineteenth century. A third proxy for size is provided by recorded numbers of full-time geographers.

We can illustrate this growth by reference to a single country (see figure 7.2). Thus for the United Kingdom, a striking contrast between university geography in the 1980s and 1950s is simply that of size. Today there are some forty university departments in the United Kingdom with an academic staff of around 600, graduating some 1,500 main-subject geographers each year. If we expand our definitions to include geography graduates with joint degrees and those taking courses in polytechnics, then the total begins to approach 2,000. In the immediate post-war years, the situation was complicated by returning ex-servicemen, but the best estimate available then

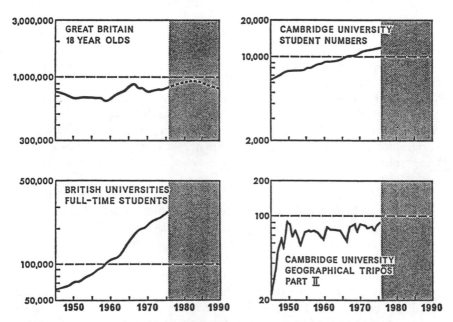

Figure 7.3 Growth of numbers in a single geography department.
Source: P. Haggett, 'Geography in a steady-state environment', *Geography*, Vol. 62 (1977), fig. 1, p. 160. Data for the fourth figure refer to Part II of the Geographical Tripos and were kindly supplied by Dr D. R. Stoddart from his extensive files on the history of the department.

places the number of graduates at less than one-third of the current numbers. Measured over the longer term, the graduate population has been doubling every twelve years for the last half-century, growth being below that rate before World War II and above that rate since. However, as figure 7.3 shows for Cambridge, the growth for individual departments has been less rapid and expansion has stemmed more from the creation of new departments than from the exponential growth in existing ones.

IMPLICATIONS OF HIGH GROWTH

For geography, as for any discipline showing rapid growth, there are important side effects. These are seen in five ways: in the age structure of practitioners; in the tendency to fission; in ephemeralization of literature; in institutionalization; and in international communication.

The most obvious implication has been a progressive change in the age structure of the profession, with the 'birth rate' of new (and generally young) entrants into the profession exceeding the 'death rate' by retirements and resignations. Geography has therefore become increasingly youthful; most geographers who have ever practised as university geographers are still practising now.

Size has also brought fission. With the increasing sizes of departments has come the tendency to split into smaller and smaller sub-groups. The term 'geographer' is increasingly being replaced by a qualifying adjective; geographers identify themselves by labels such as 'biogeography' or 'South Asian geography'. For example, I hold a chair of Urban and Regional Geography at Bristol, so called to distinguish it from the chair of Physical Geography; when there was only one chair, it was simply a chair of Geography.

Rapid growth also has an impact on the literature of geography. The rapidly expanding volume of new books and papers means that each new contribution has a shorter and shorter place in the sun as 'the most recent statement' before it is trumped by three others. In this quickening tempo, research findings tend to become ephemeral and retain their currency – if not their intrinsic validity – for ever shorter periods. Stoddart has demonstrated that the half-life[7] of citations in the geographical literature fell strikingly during the 1950s and 1960s: from over fifteen years at the beginning of the period to ten and then to five at the end. This high turnover mainly reflects increased publication rates but may be coloured by a trace of faddishness and *haute couture* more akin to a fast-growing fashion industry than to an established research field. As with papers, so with maps. The flood of new sheets has reduced the probability of any one sheet in a major map collection being consulted.

Institutionalization is a fourth implication. As geography gains in size we see the role of any one individual within the population of scholars marginally reduced. It is now less likely that any single geographer could have as dominating an effect on their period as a William Morris Davis or an Isaiah Bowman had on theirs. The early history of academic geography in both Australia and Canada was strongly shaped by the personality of one man, Griffith Taylor (plate 22). Parsons says of Sauer that he 'towered like a Chimborazo over the field of academic geography'.[8] The argument is not that an individual geographer today is less gifted or less driven by the work ethic than his predecessors; that is a judgement for future historians. It is that he or she is a much smaller part of a much larger profession and, more importantly, that the profession has become much more institutionalized.

Geographical research is now more corporate in character. A research inquiry may now regularly involve many scholars (sometimes from different disciplines), more than one institution, funded from both internal university funds and external foundations. Thus my own work on the diffusion of epidemics has been part of a group effort, involving eight colleagues in different institutions with disciplines running from virology through computer science to the Icelandic language.[9] Without this co-operation the research would have been impossible. The infrastructure of an institution – its laboratories and libraries, its computer facilities, its technical assistants and the presence of eminent scholars in related fields – are all now critical to most research endeavours. Research funds are generally given to institutions, not to individuals so that an institution's ability to attract research funds and the grant-attracting status of an individual scholar are critically interrelated in a symbiosis. Research foundations expect an applicant to be working in a strong and well-supported base, the 'well found laboratory' of the Research Council grant regulations.[10] Philosophical and theoretical work and study using archival sources or personal field data are still important, and here the institutional environment is less critical.

Finally, the communication channels of geography are becoming more international. In the period before World War II the discipline was dominated, both intellectually and numerically, by the ideas of German and (to a lesser but still significant extent) French geographers. The last quarter-century has been marked by a change in that position through the increasing dominance of English-language journals and publications, reflecting both the importance of North American research and the increased role of English as an international language of scientific communication. Such global convergence of geographical work occurs in the increasing use of mathematical languages for communication. It remains a moot point whether the increasingly mid-Atlantic flavour of much geographical research is an advantage, or whether the more distinctive national schools of earlier decades preserved a valuable flexibility of emphasis and approach. Yli-Jokipii shows how Finnish geography between 1920 and 1980 developed in a rather independent fashion, with its external sources of ideas (mainly coming from Germany) being cut off by wartime barriers.[11] The value of independent thinking is shown in many novel Finnish approaches to problem-solving in both physical and human geography, as is shown in papers in the three leading Helsinki-based journals, *Fennia, Terra* and *Acta Geographica*. In a converging world, isolation has its value.

LIMITS TO GROWTH?

The fact that rapid growth will be self-limiting was argued by Sir George Thomson in his book *The Foreseeable Future*, first published some thirty years ago.[12] In reviewing the technical changes over the last 300 years, Thomson pondered on the likely future rates of material progress. The question is familiar enough, but what makes Thomson's book of special interest is that he then goes on to lay down some of those ground rules on which his question may be answered. For example, one such ground rule is provided by what he terms those 'principles of impotence' which identify things which cannot be done.

Exponential growth illustrates Thomson's principles. Let us take, for example, doctorates in geography from British universities. During the 1960s the numbers of such graduates grew at a rate which implied a doubling every six years. Such exponential growth was necessarily short-term and anomalous: given that rate of increase, numbers would have leapt to 1,000 a year by the 1990s, and to 10,000 a year twenty years later. To continue with the fantasy, the number of geography doctorates would pass the forecast population of the United Kingdom within the next century!

Some of the evidence now coming to hand points to a slowdown in the growth of most universities; the growth of most geography departments has slowed until a plateau was reached around 1985.[13] Thus we might expect the average age of geographers to rise or the splitting off of individual sub-disciplines to cease. On most assumptions the number of undergraduates looks likely to remain stable unless recent initiatives for widening access are followed up. Against this stability, research funding is racing ahead still and the number of research staff is growing rapidly. This is reflected in the graphs showing changes in the academic staff in UK universities plotted in figure 7.2.

CYCLES AND FASHIONS

Most changes discussed in the preceding section are monotonically moving in a single direction (at least over the period of interest), but others swing to a high or low mark and then reverse their trend. Such short-term waves are less easy to discern than their longer-term counterparts but can be seen in three oscillations in geography: first, the oscillation between regional and systematic approaches to geography; second, between physical and human geography; and third, between local and global emphases.

Evidence for such instability can be garnered from the study of the contents of papers in leading geographical journals. Figure 7.4 shows cyclic trends in the work of both German and Finnish geographers over several decades. Diagram A shows the relative decline in physical geography being replaced by a late surge in the 1970s, as well as stronger oscillations in the balance of systematic and regional geography. Both trends are based on leading German-language journals. The second diagram, B, shows the same kind of oscillations in Finnish geography over a fifty-year period.

The apparent decline in regional geography has been a cause for concern, as we saw in chapter 4. The rapid increase in geographical research over the last generation has taken place predominantly in areas where geography appeared to overlap with other systematic disciplines. In regional geography the trend has reversed, with a smaller proportion of time being devoted to this central activity. Growth has been atoll-like, with fastest growth on the outer margins. Concern had focused on the neglect of research in those central areas which have traditionally given the discipline its over-riding coherence and purpose. This thesis was developed most ably by geographer Charles Fisher in his thoughtful paper 'Whither Regional Geography?',[14] where he observed that, while systematic geography flourished like the biblical bay tree, regional geography (however defined) appeared to be declining and even withering away. Not only was there a noticeable decrease in the importance attached to it in the university syllabus, but this process had also spread to schools.

But the decline recognized by Fisher has not been a constant trend but rather a part of a cyclical process. A decade later another regional geographer, John Paterson, could give a much more optimistic picture, pointing to the stubborn survival of regional geography and its parallels with the survival of faith.

> We are repeatedly assured, and by a wide variety of voices, that we live today in a post-Christian world. Yet individual churches remain obstinately full, presumably of those who have failed to heed the news.

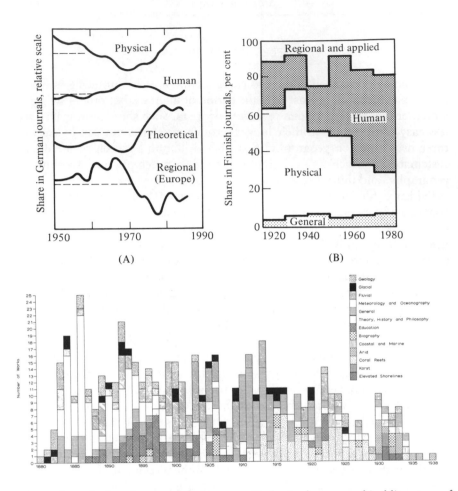

Figure 7.4 Cycles of geographical interest. Cycles in the geographical literature of European countries: (A) German-speaking countries, (B) Finland. (C) Research cycles in the lifetime of a single geographer. The number of works produced each year by William Morris Davis (1850–1934) between 1880 and 1938. Note the three waves of interest in (i) meteorology and oceanography, (ii) education, and (iii) coral reefs.

Source: (A) Elisabeth Lichtenberger, 'The German-speaking countries'. In R. J. Johnston and P. Claval, *Geography Since the Second World War: An International Survey*. Beckenham: Croom Helm, 1984, pp. 158–84. Pentti Yli-Jokipii, 'Trends in Finnish Geography in 1920–1979 in the light of the journals of the period'. *Fennia*, Vol. 160 (1982), pp. 95–193. (B) R. J. Chorley, R. P. Beckinsale and A. J. Dunn, *The History of the Study of Landforms or The Development of Geomorphology, Vol. 2. The Life and Work of William Morris Davis*. London: Methuen, 1973, Appendix IV, pp. 826–7.

Either they disagree, or they are too stupid or too stubborn to alter their habits.[15]

For evidence of today's renewed interest in regional geography one has only to scan the new titles in publishers' catalogues, or to note the heavy enrolment in university courses with a regional focus. Paterson's views are of especial interest since, unlike many commentators on regional geography, he writes as an active and experienced practitioner in the field. His *North America* now runs into its sixth edition, while his memorable lecture series on that continent, given at Cambridge in the early 1950s (and on which the first edition of his book was based), uniquely attracted undergraduates from faculties well outside geography.[16]

Other cyclical features have also had their comings and goings. The tide which ran so strongly towards human geography in the 1960s has now turned and the combined force of both environmental reality and ecological concern is enforcing growing interest in physical geography. Within human geography the contrasts between approaches at the macro- and micro-geographic levels parallel the dilemma in economics.[17] Other dichotomies relate to the swinging balance between the emphasis given to individual cases as opposed to general theory, on the relative weight attached to the regional and systematic aspects of the subject, and so on.

To view these changes as dichotomies between two alternatives is perhaps to oversimplify. The various traditions within the field recognized in chapter 1 – ecological synthesis, spatial analysis, regional synthesis – might be seen as the three corners in a standard ternary diagram (see figure 7.5). In some periods the profession appears to move away towards one corner; in others to occupy the middle ground. Of course, the use of only three co-ordinates in our system is oversimplified; we need to add at least one

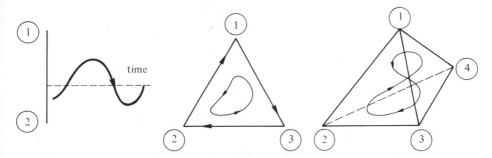

Figure 7.5 Oscillations in geographic work within a two-pole, three-pole and four-pole framework. The time track of the centre of gravity of work fluctuating between the attractors posed by research opportunities at different periods.

more, the historical time horizon, to give a fuller picture of the complexity of the field.

SHIFTS AND DISCONTINUITIES

The third type of change is the most abrupt and the most controversial both to identify and to explain. A discipline that is following a particular course may, rather abruptly, strike out in another and different direction. For example, my own undergraduate period was dominated by Richard Hartshorne's monumental *Nature of Geography*, still the most magisterial of English-language statements on the philosophy of the field.[18] By the early 1960s the position had changed and the next few years were, at least for the author, a period in which mathematics and logical positivism held great fascination. The publication of David Harvey's *Explanation in Geography* at the end of the decade may be taken as the high water mark of that movement.[19]

During the next decade any consensus was to evaporate. From within, a series of 'new' geographies came forward to replace the newly established quantitative orthodoxy. Two major contenders were humanistic geography, with new foci in phenomenology, idealism and existentialism, and radical geography, with its stress on social responsibility and policy-related research. From outside, the dominant positivist philosophy which had underlain – implicitly rather than explicitly – the approach of the spatial school came under increasingly critical review.[20]

With this joint erosion, the structure of geographical research shifted. Common bonds of morphological method shared by both physical and human geographers weakened and the division between these two sides of the subject became more noticeable, even within quantitative geography. Cut off from its older geographical roots, human geography began to drift towards the rest of the social sciences. As it did so, its individual image became more blurred and some of its practitioners argued for a redefinition of the field. The extent of the changes varied both from department to department within the United Kingdom, and, still more strikingly, within the different European countries. The publication of Derek Gregory's *Ideology, Science and Human Geography*[21] towards the end of the 1970s showed how far and how fast the position had changed since Harvey's major statement a decade before.

This compressed account gives only a caricature of a very complex pattern of change. Some departments and some specializations (e.g. historical geography) stand apart; my own country may also have been more volatile than geography in the rest of the academic world. Nonetheless, the contention that very rapid change has occurred within a major part of the discipline stands.

KUHNIAN REVOLUTIONS

But how do we explain such rapid shifts? Why should a subject suddenly regard one approach as acceptable, another impossible?

My interest in this puzzle was first aroused in the 1960s by the publication of Thomas Kuhn's *The Structure of Scientific Revolutions.*[22] Kuhn argued that the history of science was one of discrete episodes dominated by a particular mode of research and separated by briefer periods of change; in other words, that the path followed was more revolutionary than evolutionary. Central to Kuhn's ideas was the concept of the paradigm, the framework within which scientific activity proceeds. It prescribes the accepted facts, the puzzles that remain to be solved and the procedures to be used in seeking solutions. Normal science may be seen as cycles of puzzle-solving in which 'procedures' are applied to 'puzzles', adding to the stock of 'facts'. All three parts are embedded in textbooks and regarded as 'normal' for the science in question.

But not all attempts at puzzle-solving are successful. A few fail to produce any results or produce answers that do not fit readily within the pattern of accepted facts. Commonly such results are regarded as anomalies and they are set to one side while the normal cycles of puzzle-solving continue. The slow accretion of anomalies eventually reaches a sufficient volume to divert some researchers away from normal puzzle-solving to seek some alternative explanations which will encompass both the anomalies and the stock of facts established by the existing paradigm. To be successful, the new paradigms will need to change the search procedures and the definitions of the puzzles to be solved. A revolution in science occurs if this takes place, the new paradigm replaces the old, and a set of new texts and courses establishes its scholarly dominance until the next shift.

The ideas of Kuhn's first edition were seized on by geographers in both North America and Europe, and used to interpret the recent changes in both physical and human geography (e.g. the rise of quantitative geomorphology in the Columbia School headed by A. N. Strahler and of quantitative economic geography in the Washington School headed by W. L. Garrison). One view of this so-called 'quantitative revolution' is given in figure 7.6. My own attention was drawn to Kuhn's book by David Stoddart at Cambridge. The ideas were incorporated in the Madingley Lectures and eventually found their way into the opening chapter written with Richard Chorley for

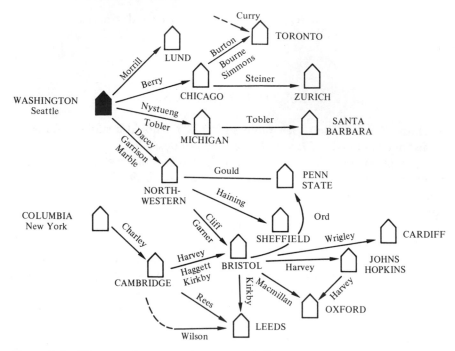

Figure 7.6   The quantitative revolution as a diffusion process. A highly simplified and incomplete picture of some of the moves of geographers from two leading United States graduate schools in the 1950s and 1960s, and their impacts on the United Kingdom. Some of the second- and third-order moves occurred after 1970. For simplicity only one centre in human geography (Washington) and one in physical geography (Columbia) have been retained.
Source: P. Haggett, 'Revolutions and quantitative geography: some personal reflections on the bicentennial'. Paper to the Sixth European Colloquium on Theoretical and Quantitative Geography, Chantilly, France, 6 September 1989.

Models in Geography.[23] From there they had wide circulation within geographical communities.

I see such Kuhnian shifts as a trade-off between fidelity and complexity. Fidelity represents the extent to which an idea or model faithfully describes the real world; complexity measures the degree of investment we need to understand the idea or model which is being proposed.[24] The relation between the two is set up in figure 7.7 with fidelity on the vertical axis plotted against complexity on the horizontal. Ideas with the greatest appeal are those which are powerful and simply expressed (for instance, laws of gravitational attractions or the relations between energy and mass). Conversely, ideas with low utility are both complex and difficult to master, yet

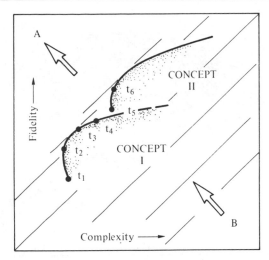

Figure 7.7   The trade-off between complexity and fidelity in model building. A possible model for shifts from one conceptual framework (I) to another (II). The diagonal lines indicate indifference contours on a fidelity–complexity surface. Time period t(5) indicates the shift point between the two frameworks.

explain little of the human world. It is possible to add to figure 7.7 a parallel series of contours representing lines of equal utility running between these two extremes.

The progression of a model may now be seen as a trajectory. If we follow the course of model A it is introduced as a_1. Refinements push it further up the contour field until a_3; later developments (a_4, a_5) add to the complexity at a higher rate than additions to fidelity so it slips back along the contour field. The widening gap between the later refinement of the model and the model at its peak performance (a_3) is shown by the stippled zone. As this increases, so pressure grows for the introduction of an alternative model, B. In its initial stage (b_1) the replacement model may not be as useful as the 'old' at its best (a_3): it is, however, seen to have a potential for development which will eventually carry it further up the slope.

In physical geography, the progressive modification of the Davisian cycle of landscape development (figure 7.8) since its first statement in 1880 may be seen as one example. The idea was powerful in its simplest form when applied to humid landforms. As it was extended to take in a wide range of global conditions – arid, semi-arid, peri-glacial – so it became overly complex and less efficient. In settlement geography, the progressive modifications of the Christaller model by Lösch, and later Dacey, might be viewed as another example.

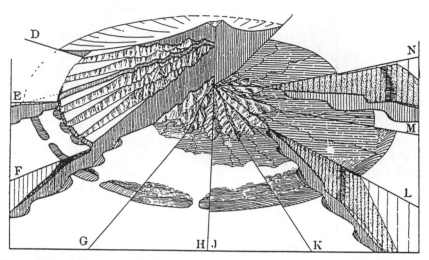

Figure 7.8   Adding complexities to geographic models. The Davisian model of landform development was originally applied to river-eroded landscapes. Extension to other landscapes extended its range but only at the cost of adding significantly to its complexity. Davis's ideas on a cycle of coral island forms is shown.
Source: R. J. Chorley, R. P. Beckinsale and A. J. Dunn, *The History of the Study of Landforms or The Development of Geomorphology, Vol. 2. The Life and Work of William Morris Davis.* London: Methuen, 1973.

THOMIAN BIFURCATIONS

Kuhn's model came under increasing criticism during the 1970s and he himself revised his schema. But the dilemma of rapid shift remained and attention turned to alternative explanations. A second approach was to be provided by the French mathematician René Thom who saw abrupt change in the natural world as explicable in terms of catastrophe theory.[25] By 'catastrophe' Thom does not mean 'disaster', but simply sudden change. Thom was fascinated by the way small changes in a 'control' may bring sudden switches in 'response'. For example, if we gradually lower the temperature of water there is a point at which there is a sudden switch in the response state, from water to ice. If confidence falls slowly on Wall Street, there may be a sudden loss of nerve in which the bottom abruptly falls out of the stock market. Although the mathematics of catastrophe theory is complex, it is possible to use its simpler ideas to throw light on why jumps within geography appear more characteristic than gradual, evolutionary change.

Let us begin our answer by assuming that a geographer's decision to

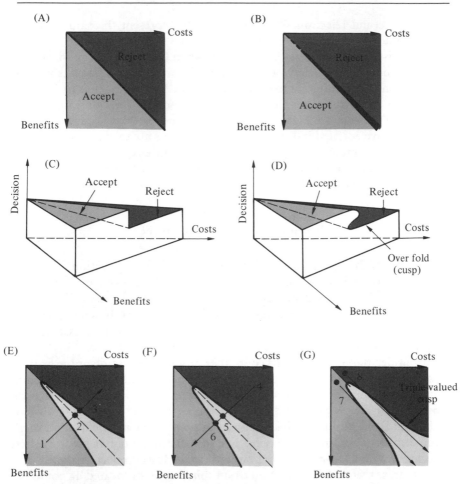

Figure 7.9 Model of paradigm shift within geography in terms of movement on a benefit–cost surface. The interpretation of the diagram is given in the text.
Source: Peter Haggett, 'Mid-term futures for geography: a framework for speculation'. *Monash Publications in Geography*, No. 16 (1977), fig. 3, p. 8.

accept or reject a given approach to the field is based on the relative balance of its perceived benefits and costs. (By benefits we mean the insights or analytic advantages that come from a particular approach.) Accepting one rather than another rests on a net balance of benefits over costs, and vice versa. This relationship is shown in the form of a graph in figure 7.9. The diagonal line in A therefore represents an indifference line where perceived costs and benefits are exactly equal; it marks the boundary between acceptance and rejection. We may go on to represent the same line in a three-dimensional version [as in C].

But this simple accept/reject line as shown in C is an oversimplification of a researcher's actual behaviour. For, where both costs and benefits are very low, we may be indifferent to each outcome, i.e. the approach is too weak to be worth either supporting or opposing. Where the stakes are raised and costs and benefits are high, we may tend to persist with a given decision once we have adopted it. Both these modifications are taken into account in the fourth diagram D by showing the acceptance/rejection decision as a surface with a symmetrical fold lying along the axis of the indifference line shown in the two preceding figures.

FOLDS AND CATASTROPHES

Adoption of this revised picture has some interesting implications. To show these, let us project the surface shown in three dimensions in A on to the two-dimensional base delimited by the costs and benefits axes. The fold in the surface now forms a triangle shaded in E. Note that the shaded triangle may now take on three different series of values since it is essentially a triple-sheeted fold in the original acceptance/rejection surface.

Consider the position of a geographer at location 1 in diagram E. If we assume benefits remain unchanged but evidence on the cost of maintaining an existing approach builds up, his position moves horizontally across the benefit/cost space. Since he is on the upper part of the fold (shown in diagram D) he changes to rejection not at point 2 when benefits are exactly equal to costs (i.e. the indifference line of diagram A) but at point 3 where he suddenly falls from the upper 'accept' to the lower 'reject' surface.

Contrast this with the position of an initial rejector located at point 4 in diagram F. In this case the assumptions are reversed and his position moves vertically over the lower surface of the fold as evidence of the benefits as accrued. Again attitude change occurs not at point 5 but at point 6 when he suddenly leaps on to the upper surface of the fold. Of course, the examples chosen are oversimplified in order to make a point; actual patterns might be expected to be much more complex.

The position of two geographers located originally at points 7 and 8 is shown in diagram G. Both costs and benefits are small, and both people may be expected to be indifferent to the positions they hold. However, if the data on costs and benefits accrue over time in the way shown by the parallel paths, we can see how our two geographers diverge. One follows the upper surface of the fold, the other the lower. Note that, although given exactly the same evidence, one geographer increasingly accepts the approach while the other increasingly rejects it. In other words, we dig in our heels to hold our initial positions and reinforce our original positions ever more strongly since there is now so much more intellectual capital (papers, books, reputation, research grants) at stake in the approach adopted. Geographers find it no easier than other folk to live with the knowledge that they backed the wrong horse.

SHIFTS: AN EXAMPLE

It may be helpful here to try to pin down these large ideas by a specific illustration of shifts in a narrower area of geographical thinking. We have already referred to the work of Torsten Hägerstrand of Lund University. To set that work in context we have to look back over a longer period of time and see the way in which one particular innovation wave, that of epidemic diseases, has been viewed. A summary of the various models is shown in outline form in figure 7.10. The growth of mathematical modelling of the diffusion of epidemic waves can be seen there in terms of six stages:[26]

First, empirical description, which was common from the seventeenth century through the work of Graunt and Petty on London mortality records. It continues through to today and lies at the heart of most subsequent modelling frameworks.

Seond, a shift to curve fitting and prediction began with the use of 'methods of differences' from 1840 to predict the later course of an epidemic given the opening stages. Increasingly sophisticated use of time series has persisted right through to today in Box-Jenkins and Kalman filter models.

Third, deterministic models came with the identification of the major components in an epidemic model in terms of susceptibles, infectives, recovereds, diffusion, birth and recovery rates. Early papers by Hamar in 1906 on measles and Ross in 1911 on malaria lead on to fundamental papers by Kermack and McKendrick in 1927 on a threshold theorem separating out the basic shift between endemic and epidemic states.

Fourth, stochastic models saw the restatement of deterministic models in probabilistic terms occurring from the 1920s. This allowed the study of small groups through chain binomial models, and the simulation of both epidemic behaviour and evaluation of different control strategies through Monte Carlo procedures.

Fifth, spatial modelling came next. Both deterministic and stochastic models were given an increasing degree of spatial sophistication from the 1940s onwards. The spread between discrete communities was studied by

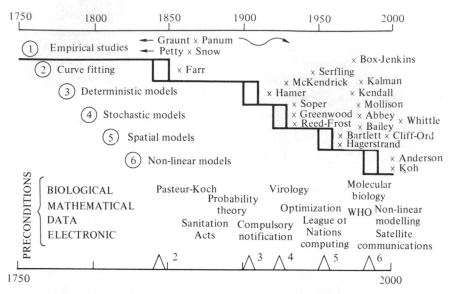

Figure 7.10 Shifts in diffusion modelling. Sequence of six shifts in the modelling of epidemics as a spatial-diffusion process. The upper part of the diagram illustrates some of the key studies in each of the six modes, while the lower identifies some of the preconditions for the modelling shifts.

Bartlett and modelling on lattices of increasing complexity has also been developed.

Sixth, came non-linear modelling with increasing interest in the 1970s in both catastrophe and chaos modelling for epidemic processes. New light is being thrown on the puzzling switch from endemic to epidemic states.

Although recorded accounts of epidemics go back to the ancient Greeks, little genuine progress in epidemiology came until the nineteenth century. What caused this delay?

Some of the reasons are suggested in the lower part of figure 7.10. Wholesale and relatively accurate recording of the weekly incidence of infectious diseases over small geographical areas did not become common in developed countries until the middle of the century, i.e. a firm spatial data base was not established. Then there were no precise hypotheses about the spread of disease suitable for expression in mathematical terms. The spectacular work of Pasteur (1822–95) and Koch (1843–1910) laid a firm bacteriological foundation and the observations of Panum on measles, Snow on cholera and Budd on typhoid each allowed the manner of geographical spread to be established. Finally, the requisite mathematical techniques and computer power necessary for model building were not yet developed.

ORDER OUT OF CHAOS

So far in this analysis of change in geography we have followed classical lines, trying to split the change into distinct stages and then trying to isolate each of them. But the real world is messy and in reality all these changes are linked together. Most of the change in geography is not orderly and stable; rather the subject bubbles and seethes with chaotic change and disorder.

This type of change fascinated the Russian-born Nobel prizewinner, Ilya Prigogine, and he has developed a theory of 'fluctuations' which links back to Thom's ideas.[27] In this Prigoginian world, all systems are made up of sub-systems which are continually fluctuating. A single fluctuation, or a combination of them, may become so powerful that it initiates a positive feedback which shatters the existing organization. At that point – the singularity or bifurcation point – it is impossible to predict in which direction the system will move. It may leap towards chaos, or leap towards a new and higher level of organization.

Prigogine's colleagues have illustrated the way in which his ideas would

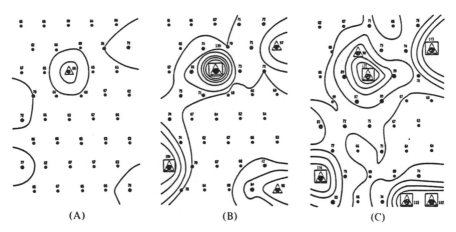

(A) (B) (C)

Figure 7.11 Order through fluctuations. Prigogine's ideas of order through fluctuations as applied to the spatial evolution of an urban system. The original Christaller-like set of settlements is subjected to very small shocks which leads to a non-symmetric growth of urban functions.

Source: P. M. Allen and M. Sanglier, 'Urban evolution, self-organization and decision-making'. *Environment and Planning A*, Vol. 13 (1981), figs 4–7, pp. 171–5.

modify the ordered view of human settlements embedded in the Christaller and Lösch models. These form hexagonal networks with the most important towns at the intersection points and smaller centres arranged in molecular-like symmetry around them. (Check back to figure 4.1 for a diagram of the Lösch model.) Figure 7.11 shows the first stages in a regional history of urbanization. This is based on the same reasonable rules of economic behaviour that are inherent in the classical models, but the launching points for new enterprises have random (chaotic) elements built into them. The different computer simulations are conducted on Prigoginian lines and capture the patterns of growth and decay, capture and domination, opportunities and alternatives which we know characterize actual spatial histories. Chance nudges the spatial pattern into a unique configuration. A whole set of new models of this kind is now being developed under the general theme of 'chaos science'.[28]

The world's geography is strewn with the litter of such place-bound memories so that there is never an initial plane on which to start our equilibrium models running. In Prigogine's words: 'we live in a world where different interlocked times and the fossils of many pasts coexist.'[29] Past palaeoclimates of geographic thought[30] will continue to nudge the present structure of the subject in unexpected directions. Our past clatters along behind us like a toy duck on a string.

8

Geography Future

'Are we nearly there?' Alice managed to pant out at last. 'Nearly there!' the Queen repeated. 'Why, we passed it ten minutes ago! Faster!'
Lewis Carroll, *Through the Looking Glass and What Alice Found There* (1872)

Like individuals, university departments have birthday parties at which various 'rites of passage' are marked. The Geography Department at Cambridge University celebrated its centenary in July 1988 and I found myself following David Stoddart of Berkeley and Richard Chorley of Cambridge, in speaking respectively on the past, the present and the future of Cambridge geography.[1] Such enjoyable events share some of the characteristics of Greek theatre, a Barnum and Bailey circus and an English garden party. Speaking last, you always find that your favourite foxes have been already shot, but that is compensated by the fact that the last period – the future – is uncertain, fluid and capable of being shaped in quite unexpected ways.

Before writing my notes for the occasion, I looked at a score of books and papers written in the 1970s, some modest, some strident, which told us what the next decade would hold and what our problems would be. They make discouraging reading today. None of them forecast, even as late as 1979, so much that has actually happened. The tidal wave of AIDS cases in central Africa and our western cities; the rise of *perestroika* and *glasnost* in Soviet Russia; the October 1987 crash of international stock markets and the fact that such an event did not trigger the expected slump; the roaring rise of the Newly Industrializing Countries (the NICs), particularly those around the Pacific rim; the concern over greenhouse effects and ozone layers; the superabundance of grain produced by Western Europe so that England and Wales now grow more wheat than the whole of Australia. And so on, and so on. They give us no confidence that we can guess at the next ten years, still less the next hundred.

The geographical literature is also strewn with the spent cartridges of old projections and any survey of past forecasts of future spatial patterns shows a similar chastening mixture of delusions and insights. One example of the former is from the early eighteenth century, when Montesquieu had no doubts about the future level of world population:

> After performing the most exact calculation possible in this sort of matter, I have found that there is scarcely one tenth as many people on the earth as in ancient times. What is surprising is that the population of the earth decreases every day, and if this continues, in another ten centuries the earth will be nothing but a desert.[2]

Slightly greater success has been achieved at the level of the single country rather than for the whole planet. For the United States, Brian Berry found that the surmises of Robert Vaughan in 1843 and H. G. Wells in 1902 were disturbingly accurate in seeing the general pattern of geographical change, if not its timing.[3]

THE FORECASTERS' DILEMMA

Lack of past success is one of the most powerful reasons stopping us looking forward. But there are others. First, as we saw in the last chapter, future growth is inherently unforecastable in that it lacks stable and repetitive structures. While trends and cycles might just be recognizable, shifts are not. Second, past records of attempts to look forward suggest that we are too cautious. Failure to predict the scale and nature of change (what the statistician calls Type I errors) usually outranks the prediction of events that failed to occur (Type II errors). Thirdly, if forecasts are taken seriously then they may become self-fulfilling as we seek to bend events to avoid a future we dislike or to attain a future we seek. All three factors confirm the future as essentially unknowable. To try to make sense of future events is to appear like a dog walking on its hind legs: as Dr Johnson commented, the wonder is not that it is not done well; the wonder is that it is done at all.[4]

And yet, despite all the difficulties, it seems appropriate, even inevitable, at the end of this small volume that the last essay should be on geography future. The need to look ahead remains critical. Most of the actions we are now taking in terms of investment in research, teaching and organization in geography are based on some implicit view of its future. In most cases our view of the future is intuitive and unspecific, and one of our tasks in looking towards the future of the subject is to try to make more specific the particular assumptions that underlie our expectations.

In practical terms, forecasting strikes a balance between benefits and

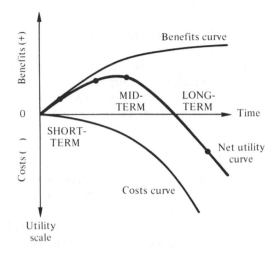

Figure 8.1 Benefits and costs in short-, medium- and long-term forecasting. Both errors and possible gains change in different ways with the increasing timespan of a forecast. The medium-term is where the net utility seems to approach its maximum.

errors. Figure 8.1 tries to illustrate how these two interact. The further ahead in time over which a method allows us to forecast, the greater its value, and this is shown by the upper curve. But this relationship is not linear over time. Its usefulness will be progressively limited since long-term projections in one sector will outrun our forecasting ability in another. Also very distant forecasts raise both practical and ethical considerations about one generation making decisions for others far ahead. The curve of error (lower curve) is much simpler. We know that, as we try to project further ahead, so the probable error band flares out in a cone of uncertainty.

As figure 8.1 shows, the joint utility of a forecast is a compromise between this convex utility and the concave curve of the likely error. Clearly, there must be a point at which the projection in time is at an optimum; beyond this optimum there is a progressive fall-off in utility until a second point is reached where errors begin to outweigh any benefits. Peak utility is gained with medium-term forecasts.

It is easier to outline the theoretical form of such a utility curve than it is to evaluate it in practice. Where detailed studies have been carried out, the forward throw of accurate forecasts tends to be very limited. For example, five-day forecasts of temperature in the United States are wrong about 30 per cent of the time, monthly forecasts 40 per cent, and seasonal forecasts nearly one-half of the time.[5]

STEADY-STATE RESOURCES

Despite this uncertainty, we can see some trends in the universities which suggest the context within which geography departments, like others, will have to operate in coming decades. The first is the resource environment, the second the technical environment, and the third the philosophical environment.

The first of these is the switch towards a steady state in scientific funding. To put it in terms of a Malthusian equation, scientific research tends to grow exponentially, while the world's gross global product grows linearly – or, if not linearly, then at a lower rate. For a brief period in the 1960s the two trends could be matched, but since then the gap has widened hugely. From time to time, in subject areas, Government may find funds to rebuild a laboratory or two. But for most of the near and mid future we shall have to design programmes knowing that the resources available are not there, or at least not there on demand from central funds. So, for geography departments too, there will be hard choices on which fields to till and which to leave fallow or to support from outside funds.

My own period of service on the relevant UK Government committee coincided with this switch and saw a major revolution in the funding of British universities.[6] Since its establishment by the Treasury in 1919 to its abolition seventy years later, the University Grants Committee (UGC) had the broad task of advising the Government of the day on the financial needs of universities. Its terms of reference were:

> To enquire into the financial needs of University education in the United Kingdom; to advise the Government as to the application of any grants made by Parliament towards meeting them; to collect, examine and make available information relating to University education throughout the United Kingdom; and to assist, in consultation with the Universities and other bodies concerned, the preparation and execution of such plans for the development of the Universities as may from time to time be required in order to ensure that they are fully adequate to national needs.[7]

The financial cost of supporting universities in the pre-war period was a relatively tiny one but by the mid-1980s the annual figure had reached over £2 billion. Even for a relatively small subject like geography, with some 7,000 students in 37 universities, the cost was over £20 million. The annual cost of a large geography department is now over £0.8 million. Like all other forms of Government spending, university funding was subjected to constraints and cutbacks and by the mid-1980s it was clear that new systems of funding were going to be needed. Figure 8.2 shows the costs of geography in British universities as compared to other major subjects.

The details of the new funding formulae and their application are now part of history, well-known to British academics and wearisome for overseas readers.[8] What is significant for this essay is that the British exercise is not unique. Parallel thinking is taking place in Australia and New Zealand, in France and Sweden, in Japan and Indonesia. In many parts of the world, the long exponential growth of university funding is coming to an end, with profound implications for university geography.

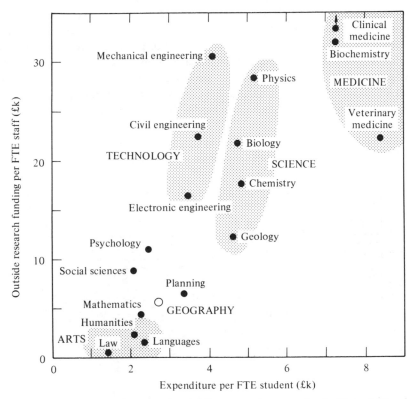

Figure 8.2   The costs of geography I. The costs of geography in United Kingdom universities in the middle 1980s in terms of expenditure per student and outside research funding. The transitional nature of the subject in respect to the humanities and sciences is shown.

Source: Universities' Statistical Record, *University Statistics 1986–87. Vol. III Finance.* Cheltenham: University Grants Committee, 1988.

Reactions to the steady state have taken a number of forms. Centrally, an increasing number of academic subjects in British universities are being subjected to rationalization exercises. These range from the study of minority languages through the social sciences to heavy sciences like physics, chemistry and geology.[9] Even where such rationalization programmes are not operating, funding formulae now contain elements of judgement on the research performance of a department. Because of the difficulty of measuring research on a 'judgemental research scale' (JR) it would be inappropriate to refer specifically to geography. There was however a high degree of correlation between the JR rankings for whole institutions.

In a controversial document,[10] the UK's senior research council body suggested a three-tier system for funding universities, based on whether they were research-orientated, teaching-orientated or an admixture of the two. Although the scheme met general hostility on the grounds that teaching and research were inseparable, the concentration of research council funding in the leading institutions is in reality already high. I cannot speak for other disciplines, but for geography the proposed split between research and teaching is profoundly misguided. Courses in regional geography by university dons who have not immersed themselves in that region are surely a recipe for disaster.

This does not mean to say that costs can be ignored. The detailed costs for British geography departments are given in figure 8.3. Diagram A shows the

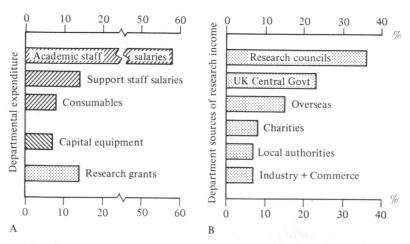

Figure 8.3 The costs of geography II. Breakdown of the costs of a typical United Kingdom geography department in the middle 1980s. (Left) Departmental expenditure. (Right) Sources of research funding.
Source: Universities' Statistical Record, *University Statistics 1986–87. Vol. III Finance.* Cheltenham: University Grants Committee, 1988.

total recurrent costs and B the sources of research funds.

In a period of low growth, departments are likely to have to find additional research funds from outside sources and to show increased specialization at graduate, and possibly even undergraduate, levels.[11] Both results are already happening. This is starting to involve rationalization of the use of resources to emphasize particular features – for example, costly equipment at one, specialized, regional library and archives at another, and joint-degree specializations at yet another. In the United Kingdom, such rationalization has already occurred in terms of the recognition of 'centres of excellence' (for instance, in Asian and African studies). The policy of grant-giving councils in making graduate studentships available only at departments with established reputations for research in environmental fields is being considered by other research councils of interest to geography.

It appears likely that, in my own country at least, we are moving towards a situation where geography departments become more different from one another, at least after the basic undergraduate course requirements of the first one or two years have been met. Taking a Darwinian view of the University system, one can hope that those departments with a successful combination of courses and research will multiply. But in this highly competitive world of gainers and losers it would be a brave forecaster who could swear that the first will compensate for the second.

A CHANGING TECHNICAL ENVIRONMENT

A second factor relates to technical change. Here the high equipment costs of some research, like the capitalization of farming, affects the organization of departments. Just as the introduction of the McCormick reaper in 1831 allowed new and larger fields to be tilled, and made some farming less economic, so the impact of the electronic revolution (environmental remote sensing, massive spatial data files and computer mapping) affects geographic cultivation (plate 23). In this hi-tech environment it is easy to forget that universities are in the business of long-term and sustained-yield farming, not training academic tractor drivers or taking a quick cash crop from the latest Government handout or company contract. The twin challenges of rapid scientific change and greater public accountability bring new pressures on the geographical community.

The convention at Bristol University, like many others, is that a newly appointed professor has to give an inaugural lecture. So back in 1966 I cast around for an appropriate theme which might interest fellow academics in the other faculties and came up with 'Geographical research in a computer environment'.[12] At that time the university had two very elementary computers, an IBM 1620 and an Elliott 503, and punched cards, paper tape and batch processing were the order of the day. Computing was a laborious and time-consuming business which interested only a minority of the university. I had nothing very profound to say and the event passed off with polite comment: 'gently radical with some interesting slides' was the best summary of reactions.

Looking back nearly a quarter of a century later, I see that my fault was not in being radical (however gently), but in not being radical enough. The technical changes have been profound. We now have, just in my own medium-sized geography department, more computing power than was then available to the whole of British universities. I type these pages at a desk into a small machine that can call through the Joint Academic Network (JANET) on any other system in a UK university, and with only slightly less difficulty can draw on international programs and data banks. No staff member or graduate student is more than an arm's reach away from a terminal, and few field parties do not carry massive computer power in a compact computer or through modem connections.

KON-TIKI VOYAGES

Let me give three examples of geographical work that is utterly dependent on the technical changes of the last few decades. First, in historical geography a group at London University tested the probability of ancient inter-island voyaging by Polynesians in the Pacific.[13] When Thor Heyerdahl undertook his now historic voyage on the raft *Kon-Tiki* from the coast of Peru to the Tuamotu Islands, he was carrying out a single experiment, testing whether it was possible to cross the Pacific in such a craft. To analyse thoroughly the probabilities of contact between South America and different island groups by this means would require too many voyages to be feasible. When direct experiments prove too costly, too risky or unlikely for other reasons, we may be able to turn to computer simulation for answers to our questions.

To test the probability of inter-island contact as a result of accidental drifting, the London group constructed a computer simulation model for the drift process. There were four main elements in the model: the relative probabilities of wind strength and direction for each month, and of current strength and direction of each five-degree square of latitude and longitude in the Pacific Ocean study area; the positions of all the islands and land masses in the study area, together with their sighting radius; the estimated distances that would be covered by ships given various combinations of wind and current strength; and the relative probabilities of survival of ships during certain periods at sea. Sighting radius (the distance out to sea from which islands can be seen) was built into the model on the assumption that, once land had been sighted, a landfall could be made.

As figure 8.4 shows, during each daily cycle voyages are started from given hearth areas like the coast of Peru, and a simulated course is followed until it ends in either a landfall or the death of the voyagers. By simulating hundreds of voyages from each starting point, the relative probability of contact with different island groups as a potential contact field can be mapped. The number of calculations needed to do this runs into billions.

The simulation program has already been run for various Pacific Island groups and for locations on the coasts of South America and New Zealand. Preliminary results indicate that the probabilities on inter-island links from the accidental drift of ships differ from one area to another. Wind and current patterns create environmental boundaries which make drifting in certain directions highly unlikely. Some of these boundaries coincide with long-standing anthropological breaks in the geographical pattern of ethnicity and culture such as that separating the Micronesian people of the Gilbert Islands and the Polynesian inhabitants of the Ellice Islands. Other low probabilities of contact coincide with important linguistic boundaries.

The most crucial conclusion of the simulation model was that accidental voyaging was a wholly adequate explanation of Polynesian island colonization. Over the limits imposed by the radio-carbon dates, it could account for the spread of the population from a hearth in the western Pacific to the most distant outposts – Hawaii, Pitcairn, New Zealand. Notions of superhuman navigational skills or South American origins, however romantic, were shown not to be needed.

LEUKAEMIA TO LOCUSTS

A second example of technology-dependent research is the work by Stanley Openshaw at the University of Newcastle geography department on the spatial problem of leukaemia clusters.[14] We saw in chapter 2 the way paired comparisons were used to throw light on the possible risks which nuclear power stations might pose to leukaemia cases. Much of the debate centred on the existence or non-existence of 'clusters' of cases. Were these significant or simply random? Openshaw used the full power of modern computing to test billions of possible distribution maps to check on the assumption. He found raised levels near one station but, more important, a very strong cluster of cases in a location where no obvious source had been identified and where there was no nuclear power source nearby. New leads involving other factors are now being pursued.

Table 5. Voyage 17, Cycle 5, Experiment 8, Start from Rapa on Day 222

Day	WD	WF	CD	CF	Position	
1 ....N	5	ESE	2	−29.293	144.486
2 ....WNW	5	SE	1	−29.946	142.620
3 ....ESE	3	WSW	2	−29.583	143.313
4 ....CALM		N	2	−29.733	143.313
5 ....W	4	W	3	−29.733	141.586
6 ....NW	3	WSW	2	−30.241	140.775
7 ....W	2	ENE	1	−30.260	140.365
8 ....WSW	7	ENE	1	−29.208	137.424
9 ....SSE	5	CALM		−27.545	138.213
10 ....WSW	4	E	2	−27.085	137.132
11 ....W	3	E	1	−27.085	136.289
12 ....N	1	NW	3	−27.497	136.051
13 ....N	3	E	1	−28.297	136.107
14 ....SSE	5	SE	1	−26.599	136.930
15 ....S	2	N	1	−26.249	136.930
16 ....S	2	CALM		−25.849	136.930
17 ....SSW	5	ENE	2	−24.244	136.318
18 ....WSW	3	WSW	1	−23.957	135.457
19 ....SSW	6	WSW	2	−21.797	134.300

Landed on Island 161 (Mangareva)

(A)

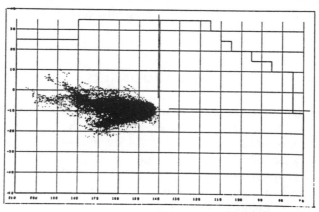

(C)

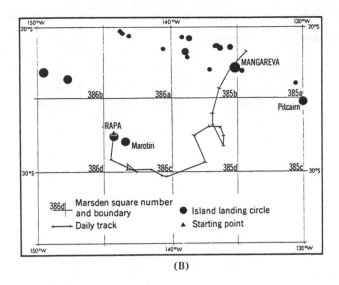

(B)

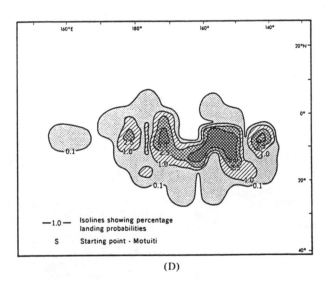

(D)

Figure 8.4 *Computer simulation of Kon-Tiki type voyages.* (A) Sample experiment from the Pacific island of Rapa and (B) its mapped equivalent. (C) The end points of hundreds of simulated voyages from Motuiti and (D) the probability field for all voyages.

Source: M. Levison, G. R. Ward and J. W. Webb, *The Settlement of Polynesia: A Computer Simulation.* Minneapolis: University of Minnesota Press, 1973, table 5 and fig. 11, p. 26; fig. 22, p. 34; fig. 14, p. 29.

A third and final example is the work of my colleague at Bristol, Eric Barrett, for the Food and Agriculture Organization (FAO) on locust predictions (see figure 8.5).[15] Since the earliest Egyptian records, locusts have been known to erupt in massive swarms, devastating crops over hundreds of square miles. Although this member of the grasshopper family lives in the arid and semi-arid lands of the world, its breeding cycle is critically dependent on rain. Rains supply the moist conditions which allow the insect to get through the forty- to one hundred-day gap between egg-laying and the adult stage when locusts then emerge in massive swarms. Locusts in swarms are both hungry and highly mobile, having been known to move distances as far as 2,000 miles in as little as three weeks when strong winds prevail.

Information about rain in the potential breeding grounds in the desert margins of the Sahara is of vital interest to the FAO locust control network. But information is hard to gain since these areas are very sparsely peopled, there are few meteorological stations, and when rains do come they fall as a random pattern of intense but highly localized storms. So studies have been initiated using environmental remote sensing. In southern Algeria and

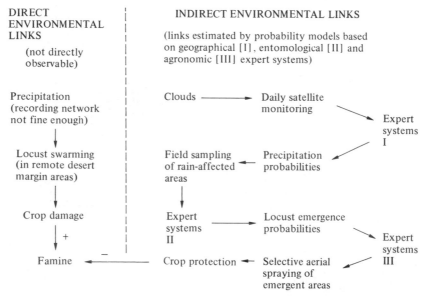

Figure 8.5   Use of satellite images in hazard prediction. Computer evaluation of satellite cloud images to predict rainfall probabilities and therefore locust dangers is part of the extensive programme of work of Dr E. C. Barrett and his colleagues in the Remote Sensing Unit, Department of Geography, Bristol University.

central Arabia satellite records from the LANDSAT series have been used to identify areas where rains are probable (based on cloud characteristics), and where rains have recently fallen (based on the green 'flush' of plant growth which comes after the rains). Once key areas have been identified from the photos, then the remote sensing team at Rome alerts the regional headquarters of the locust control programme. Control officers can then fly over the areas with light aircraft, or conduct ground surveys in the more accessible areas. If the presence of locusts in their larval stages is confirmed, control measures can be put under way. The early-warning programme is still in its infancy but this application of remote sensing promises to save millions of dollars in crop loss, always provided the political will to co-operate allows.

THE WALLS OF JERICHO

The third environment of change is an intellectual one. Even half a century ago geographers could work away at their problems and see only distantly on the horizon the far-off smoke from other academic neighbours. Today the same academic space is more heavily occupied and the campfires of near neighbours – in environmental chemistry, in engineering hydrology, in conservation economics, in area studies and the like – burn brightly around us. Many of the topics which once formed the basis of our own undergraduate courses are now matters of grave public concern; funds are there for their study if we bid for them, and some countries have whole government institutes devoted to them. Our favourite wild picnic spots are now attracting many other visitors, and partnerships have to be forged with our new and closer academic and institutional neighbours.

In this context of interchange it is tempting to argue for the sweeping away of the old disciplinary boundaries and the merging of geography with its neighbours. The argument comes from two directions. First, from those who see the intellectual ideas sweeping major areas of study (e.g. the role of social theory within the social sciences or molecular biology within the life sciences) as so powerful as to warrant major restructuring of conventional subjects. Second, from those who find the whole university system archaic, intellectually superfluous, self-indulgent and morally intolerable.[16] The first works to dismantle subjects and the second to demythologize experts. With all this noise we might well expect the Walls of Jericho to come tumbling down.

I am sympathetic to these changes in some fields but hostile to their application to geography. I hope this inconsistency is not just local conservatism. To see why it may not be, let us look back at two of the figures used in the first essay. If we look at the 'map' of disciplines surrounding geography in figure 1.3, it is easy to see our first problem. In which direction should these alliances be made? The greatest pressure for merger has come from human geographers within the social sciences, but a merger here would leave both physical geography stranded and cartography adrift, just at that time when an increasing number of our graduates are finding job opportunities in such areas as Environmental Impact Analysis or in Geographical Information Systems (GIS). To conceive of geography in terms of human geography only (important though that is) is, I think, profoundly misguided. To do so is to miss one of the essential points of the subject – whether taught at school or college, used in employment or for public education.

This dogma is made clear in Richard Hartshorne's diagram (figure 1.4). Geography's relations with other subjects are a complex intersection of planes. Our contributions lie at the interface of several scientific traditions and not within the bounds of one. That remains a central core in our contribution, and as Sauer argued, 'If we shrink the limits of geography, the greater field will still exist: it will be only our awareness that is diminished.'[17]

ACADEMIC FREE-TRADE PORTS

The tradition in our great departments has been one of co-operation. Links are forged as required. It is no accident that, when Vaughan Lewis was carrying out his Skauthoe glacial work at Cambridge in the 1950s, among those who showed interest and encouragement were physicists like John Nye, Nobel prize-winning molecular biologists like Max Perutz from the Cavendish, engineers like John McCall from Alaska and glaciologists like Maynard Miller from Seattle. Similar partnerships both within and without geography departments have been typical over the last decades with Botany, Quaternary Studies, Statistics, and so on.

Geography departments have been sheltered harbours to which foreign ships have been welcomed and from where other ships have set sail to found many departments in their own country and others. In Cambridge it was the farsightedness of members there that saw the emergence of other units within Cambridge, such as the Scott Polar Research Institute, the Cambridge Sub Department of Aerial Photography, the Centre for South Asian Studies and the Centre for Population Studies. The pressures of the work ahead will demand co-operative ventures with anthropologists and archaeologists, as well as with engineers and computer scientists. Perhaps geography departments overseas may link up in international partnerships. Most environmental and regional problems are multi-faceted and the geographic contribution, be it a major theme or a minor addenda, needs to be made in this wider research context.

I hope that, in looking outwards in research, we do not forget teaching and our links with schools. We owe an immense debt to those from our departments who have gone out to teach with enthusiasm and dedication. It is easy for those of us privileged to work in ivory towers – today 'carbon-fibre towers' might be more appropriate – to forget that the generations of bright youngsters who beat a path to these doors are not automatic or self-propelling. I very much enjoyed the great experiment in the early 1960s of the Madingley Hall lectures,[18] where teachers and university dons worked together to identify ways of introducing new ideas into geography. New journals like *Geography Review* carry on that mission.

Currently in Britain strong pressures remain on those geography places available within the university sector. Two applicants come forward for each university place in geography. Would-be veterinary students have the most difficult time in gaining places with five applicants for every place to be filled. At the other end of the scale, biochemistry has the same number of places on offer as students applying. Geography remains marginally the best placed of the non-professional subjects, but behind professional fields such as medicine, architecture, law and civil engineering. But geography is balanced dynamically between major forces of demand and supply. As Galsworthy reminds us, the status quo is, of all states, the most likely to depart.[19]

SEVEN RESEARCH AREAS

Given these resourcing, technical and intellectual changes, in which areas are the major research breakthroughs of the next decade likely to come? Let me be purely speculative and take just seven of those where I hope significant advances will be made, which will either involve geographers or affect geography, or both. The reader will need to allow for an optimistic temperament.

First, mapping is likely to change from a ROM to a RAM format.[20] Map production is conventionally geared to a final printed sheet which displays information in a fixed format; in the jargon of computer technology a map library with all its printed sheets might be likened to a 'read-only memory' (ROM). But the need for maps is often for selective, personalized data, that may be needed only for a brief period to answer a specific question. When we sketch on a memo pad a map for a newcomer to find his way to our home for supper, the resultant 'map' is cheap, quickly produced, single-purpose, area-specific and disposable (indeed we would be embarrassed were it to be kept!). Changes in computer storage already allow mapping to change to that interactive, on-demand, form within the next decade. In computer language the 'random-access memory' (RAM) is increasingly likely to replace the fixed ROMS. Selection of area, resolution or scale, and content to be displayed is already available on military systems, and the main barrier to wider dissemination is likely to be cost. The fall in the cost of both storage and purchasing should ensure that this barrier is steadily reduced.

A second breakthrough could come from a general paradigm for climatic evolution developed to bridge the gap between detailed meteorological studies (at the micro-level), geophysical models of the earth's atmosphere, and Pleistocene history (both at the macro-level).[21] This should allow some return to an integrated systematic picture of the earth's surface, welding both biological history and palaeoclimatic history in a new synthesis for our understanding of man's physical environment.

Third, a new regional geography will evolve which will supplement traditional skills by integrating quantitative models from economic geography, regional economics and regional science.[22] I have argued in chapter 4 that one of the main reasons for the quantitative revolution of the 1960s centring on systematic fields was that the methods then available were unable to handle regional complexity. Although the problem remains formidable, the development of spatial impact and regional interaction models allows a start to be made on designing a new generation of regional geographies with some additional capacity for forecasting regional change.

Fourth, the gap between micro- and macro-studies of the historical evolutions of the population of developed countries may be closed.[23] Geographers have conventionally been dogged by the problem of linking their findings at different scales, and nowhere has this been more evident than in population geography. Work during the last two decades on individual migration pathways at one scale and on migration modelling at another promises a fuller and more personalized view of migration.

Fifth, a philosophical basis for geography is slowly being worked out to encompass the existing ideas of the nature of geography within the broader base of a man-land ethic.[24] The questions of how man relates to the natural world and the cosmic order have been asked by philosophers and theologians down the centuries. The impact of the environmental crisis has brought a new interest by philosophers in geography and by geographers in philosophy. This convergence is leading to a re-evaluation of the historical evidence showing how people of different faiths (and therefore of different conservation ethics) have used resources in past periods.

Sixth, a global land-use inventory will be completed and maintained using remote-sensing data and an international network of bench-mark sites.[25] This will serve as a marker for measuring both future and past change. One of the duller but necessary jobs for geographers to finish is the establishment of international measurement standards for metropolitan areas.

Seventh, a space-proofed set of statistical tests is being established which can give bias-free estimates when analysing geographical data.[26] Analysis of time series on a secure basis was begun in the 1920s but, for spatial series, serious problems of bias and error remain to be cleared up.

Other geographers would make other predictions. But all would agree that there is no lack of major geographical problems that can soon be solved given a reasonable input of resources. Most of the ways forward depend critically on advances being made in other areas. For geography to be able to put its piece in the jigsaw puzzle, it will need other pieces to be placed in position by others. Most of the advances expected depend on research activity in microcomputers and information processing, in genetics and biochemistry, in remote sensing and space exploration. Within the next decade, entirely new possibilities may have revealed themselves which will allow new phases of geographic work, not yet foreseen, to go forward.

ON BEING UNSPECIALIZED

Although I met Carl Sauer on a number of occasions while at Berkeley in 1962, I recall some fragments of one long smoke-filled conversation in his office with particular sharpness. I had just come back from Brazil and, while there, had collected maize cobs from small markets that I had visited in the north-east for one of his old friends, Edgar Anderson, botanist and curator of the Missouri Botanical Gardens.[27] Maize was one of Sauer's many interests and he talked on that and tobacco for the next two hours. He was concerned that the short-term success of maize hybrids might force out local varieties with the long-term loss of genetic diversity, but took comfort in the fact that the less-specialized of the world's plants had proved great survivors.

This theme of the advantage of being unspecialized was one of his favourites and he had expanded on it at length in his presidential address to the Association of American Geographers in 1956 on 'The education of a geographer':

> We professionals exist not because we have discovered a line of inquiry or even own a special technique but because men have always needed, gathered and classified geographic knowledge. . . . In a time of exceedingly great increase of knowledge and of techniques we remain

in a measure undelimited and unreduced to a specific discipline. This, I think, is our nature and our destiny, this our present weakness and potential strength. . . . We welcome whatever work is competent from whatever source, and claim no proprietory rights. In the history of life the less specialized forms have tended to survive and flourish, whereas the functionally self-limiting types have become fossils.[28]

Although I think Sauer was wrong in respect of 'special technique' (surely, cartography is highly specialized?), his views need restating in the last decade of this century because so many university subjects represent a paradox. The level of specialization within any discipline is steadily increasing while the world into which students move after graduation is demanding ever more generalist and integrating skills.

A few years ago my wife Brenda and I attended a graduation ceremony in the London teaching hospital where one of our sons was studying. I recall the Dean of Medicine's valedictory words: 'You young doctors have worked hard to master your skills . . . but in ten years' time half of what you have been taught will probably be wrong . . . but, dammit, right now we don't know which half.' Which half of geographic knowledge will we have to discard by this century end?

Some observers argue that we shall probably see a decreased emphasis on undergraduate vocational training in the next twenty-five years.[29] The reason advanced for this apparent paradox is that well over half the students graduating from colleges in advanced countries enter professions for which they have received no professional preparation. This situation is reinforced by the fact that the exponential rate in the expansion of knowledge means that it is increasingly difficult to master even a significant minority of knowledge in any given academic field in an undergraduate degree course; as I argued earlier, even if such knowledge were mastered, graduates would be likely to find that it was rapidly made obsolescent by contemporary research.

We often worry, and rightly so, about the content of our geographical syllabuses and the fences around our domain of study. We appoint staff to cover this and cover that. And yet we know that content and containment are not what university education is all about. It is the flame that is lit when the brightest young minds (which universities can and will continue to attract) and bright dons (which universities can, by and large, still attract but may find it harder and harder to hold) come together. But it is not the details of equations or documents, maps or manuscripts, that remain. All these will change. The habits of geographic thought, the ways of environmental analysis, the skills of map dissection, the melding of different regional materials, the challenge of field problems – these are things of lasting value and still yield their benefit long after details of lectures are forgotten.

THE GRADUATE

Employment for British geography graduates for much of the post-war period was broadly fixed. About one-third went on to become geography teachers in schools. Another third moved into geography-related jobs: in research, in planning and in higher education. The remaining third went into general business and industry. More recently the pattern has shifted, with a decline in the proportion of graduates going into teaching and a compensating rise in the third sector. The changing techniques of industry and commercial life dictate that the present need remains dominantly for persons who are able and willing to adapt their own skills, building on their graduate education, rather than simply pursuing their original degree speciality.

What happens to geography graduates can be told in terms of maps (see figure 8.6). But perhaps a better indicator is simply to recall the six

Figure 8.6 Location of geography graduates. Distribution of the latest known addresses of 1,147 geography alumni from the leading west-coast geography department at the University of California at Berkeley.

Source: Based on maps in *Berkeley Geography 1988.* Berkeley: University of California, Department of Geography, 1988, pp. 20–1.

undergraduate colleagues who mapped with me in Portugal the area shown in figure 2.1. The leader of the party went on to join a large food company, Cadbury's, and rose from running a biscuit factory to the higher rungs of multinational management. Two others initially went into schoolteaching: one stayed and moved up into educational administration while the other won a safe seat in East London for the Labour Party in 1974 and continues as a Member of Parliament. Yet another moved to the United States and is now professor of ecology at a leading mid-western university. Sadly I have lost track of the sixth but when we were last in contact he was heading for the Gilbert and Ellice Islands as a Christian missionary.

After graduation, I found that my fellow postgraduate students in geography at Cambridge in 1954 were small in number, with most working in glaciology. Some measure of the danger of geographical field work then is reflected by the fact that only one of the three went on into academic life, the other two dying in independent field accidents in different parts of the Arctic. The few human geographers followed less dangerous paths and went on into university chairs.

RARE REWARDS

Those of us who went to research and teach in geography have been especially blessed. I have had the great privilege of spending many periods overseas from the Americas (North and South), through Africa and Asia to Australasia and the Pacific. And all along these journeys I have met geographers, running soil surveys in the Pacific or reorganizing hospitals in the Gulf or computer networks in Chicago. As a graduate of this discipline you join a community (some critics would say a mafia) of geographers with links in all five continents. And what unites them is not only their professional skills but shared memories – of what the London geographer, Sir Dudley Stamp, called 'bullock-cart days and Irrawaddy nights'.[30]

Many of the lessons we draw and the friendships we make in this profession (or should it be a calling?) are precious and will remain – despite being scattered in time and space – important for the rest of our lives. These links, personal as well as professional, span the generations. My hope for the young geographers of the next generation is that their world will be as full of interest and beauty as that which my own generation has studied. Perhaps they may find time in their future busy lives to keep links with their subject roots; to wonder why we spent so much time wrestling with spatial structures; and to include such of our insights as remain valid in their own fuller explanations of the geography of their times.

For each generation of geographers will find some new set of problems of greater challenge or consequence than those that compelled their predecessors. The current issues are different from those of my generation, and rightly so, though by the turn of the wheel older issues may come back on the agenda in some future decade. This alteration of problems is all to the good, for headway on one area may slow down, waiting for solutions to be found in different areas. I suspect that as many research barriers are outflanked as are carried by frontal assault. Many problems in global

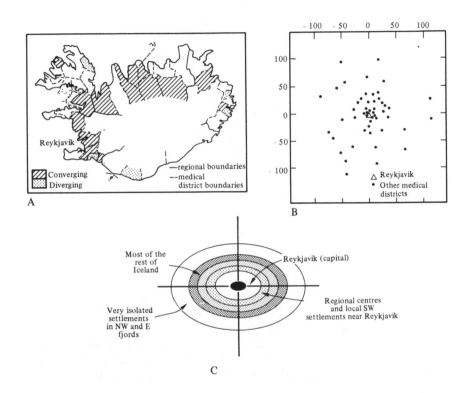

Figure 8.7   Order out of chaos. The spatial structure of epidemics in Iceland in (A) conventional geographical space and (B) in non-linear 'epidemic space'. The lower diagram (C) interprets the non-linear maps showing Iceland's capital city, Reykjavik, at the centre of a constellation of other centres with the diffusion space collapsing inwards towards the centre over time. Something that is complex in one type of space becomes beautifully simple in another, and more appropriate, type of space.

Source: Redrawn from A. D. Cliff and P. Haggett, 'Disease diffusion'. In Michael Pacione, *Medical Geography: Progress and Prospect.* London: Croom Helm, 1986; pp. 84–125, fig. 4.17, p. 122.

mapping could not have been solved, however hard we had worked, until certain advances in computing had been achieved.[31]

But for me the basic puzzle and riddles of geographical structure remain enduring. The structural symmetry of the planet as viewed from outer space, the sequence of atolls in a Pacific island chain or the terraced flights of irrigated fields on a Philippine hillside (plate 24) continue to be awesomely beautiful. When the time dimension is added, then changing structures take on added gleam. Diffusion waves become an interweaving dance of trajectories in multi-dimensional space.

This is one of the fascinations that attract young men and women into geography. To see where something fits in the spatial order of the great global jigsaw is a supreme if rare reward. Let me give you one last example. When working on epidemics in Iceland, Andrew Cliff and I found a very unsatisfactory pattern when mapping the spread vectors.[32] Putting them on a regular geographic map was accurate but the results were dull and confining. By using the scaling techniques we discussed at the start of the second chapter, we were able to recast them in a different and more appropriate spatial framework. As figure 8.7 shows, the result is simple but beautiful. All the complexities fall away and you are left with an elegant ring structure which converges over time as the diffusion process accelerates. As so often, symmetry and elegance go hand in hand with truth. Whether these middle-order spatial structures, half-way between the elegant worlds of the galaxy and the atom, will ever be shown to be what Stephen Hawking has called a dance of geometry is uncertain.[33] The superstring will certainly not stretch that far in my lifetime.

A moving section of the BBC film of the Watson–Crick–Wilkins race to unravel the structure of DNA (described in Watson's *The Double Helix*) portrays the time when physicist Rosalind Franklin comes up from King's to view the rival molecular model set up in the Cavendish Laboratory at Cambridge. She forgets her own disappointment as one of the losing team in the intricate beauty of the structure. Her 'So *that's* how it is'[34] echoes the same elation that explorers of worlds from the galactic to the sub-atomic have felt over the ages.

That there are other delights in geography – its broad acres, its siren call to distant horizons, its pervasiveness in so many current problems, its sharpening of sense of landscape and place – does not dim this simple delight in the beauty of geographical structures and the challenges posed in finding them and mapping them. Looking back into those hospital mirrors, I am glad I saw what I did and followed this exasperating but compulsive calling across so many distant horizons. I wouldn't have made a very good vet.

Notes

I resisted long and hard the notion that this book of essays needed footnotes. In the end I caved in, recalling at least one well-thumbed book on my shelves where the footnotes contain as much of interest as the text. Also, including the normal sprinkling of pitons to secure my academic ropes to the backwall of past literature also allows me to express indirectly my indebtedness to so many others for their ideas and help.

Authors are not consistent in the way they record their names, sometimes with Christian names, sometimes anglicized, and I've followed the style used on the title page in each case so there may be apparent but justified variations.

CHAPTER 1: A DISTANT MIRROR

1 Most of these essays start on an autobiographical note although this is more by accident than plan. The only published note that stitches these events together is provided by Clyde E. Browning, *Conversations with Geographers*. Chapel Hill: University of North Carolina Press, Studies in Geography No. 10, 1982, pp. 40–56. In this Browning talked with ten geographers about their 'career pathways and research styles': the ten were Barbara Borowiecki, Larry Ford, Peter Gould, Peter Haggett, Peter Hall, Melvin Marcus, Phillip Muehrcke, Harold Rose, Yi-Fu Tuan and Roger Zanarini.
2 C. A. Cotton, *The Geomorphology of New Zealand*, 3rd edn. Christchurch: Whitcomb and Tombs, 1942.
3 Barbara Tuchman, *A Distant Mirror: The Calamitous Fourteenth Century*. New York: Knopf, 1978.
4 Freeman Dyson, *Disturbing the Universe*. New York: Harper and Row, 1979, quotation from p. 193.
5 Ralph H. Brown, *Mirror for Americans: Likeness of the Eastern Seabord, 1810*. New York: American Geographical Society (Special, Publication, No. 27), 1943. Other geographers who have used the concept of mirrors are J. M. Powell, *Mirrors of the New World: Images and Image Making in the Settlement Process*. Folkestone: Archon Books, 1977; John Leighley, 'Memory as mirror'. In Anne Buttimer (ed.), *The Practice of Geography*, New York: London, 1983, pp. 60–89.
6 Alfred H. Siemens, *The Americas: A Comparative Introduction to Geography*. North Scituate, Mass.: Duxbury Press, 1977.

7   Richard Hartshorne, *The Nature of Geography: A Critical Survey of Current Thought in the Light of the Past*. Lancaster, Pa.: Association of American Geographers, 1939.

8   P. B. Medawar, *The Art of the Soluble*. New York: Barnes and Noble, 1967, quotation, p. 10.

9   William Shakespeare, *Cymbeline*, Act IV, Scene 1. The play was probably written in 1609 or 1610.

10 Hartshorne, *The Nature of Geography* (note 7), p. 249.

11 Isaiah Bowman, *The New World: Problems in Political Geography*. Yonkers on Hudson: World Book Company, 1921.

12 Letter from Morris Town, 26 January 1777. In *The Writings of George Washington, 1745–1799*. Volume 7, p. 65; cited by Isaiah Bowman, *Geography in Relation to the Social Sciences*. New York: Charles Scribner's Sons, 1934, p. 85.

13 Carl O. Sauer, 'The education of a geographer'. *Association of American Geographers, Annals*, Vol. 46 (1956), pp. 287–99, quotation from p. 289.

14 P. Haggett, 'Geography'. In R. J. Johnston, (ed.), *Dictionary of Human Geography*. Oxford: Basil Blackwell, 1981, quotation from p. 133.

15 Richard Hartshorne, *Perspectives on the Nature of Geography*. Chicago: Rand McNally, 1959, quotation from p. 21.

16 Carl O. Sauer, 'The morphology of landscape'. *University of California, Publications in Geography*, No. 2 (1925), pp. 19–53.

17 Geography is recorded in British university statistics in several ways. For financial purposes it has since 1984 formed one of 37 separate cost centres and is separately recorded in *University Statistics, Vol. 3, Finance*. Cheltenham: Universities' Statistical Record for the University Grants Committee (annual). Statistics on geography students and graduates are available from 1960 onwards, but in 1985–86 Geography was split between physical sciences and social studies in *University Statistics, Vol. 2, First Destinations of University Graduates*. Cheltenham: Universities' Statistical Record for the University Grants Committee (annual).

18 This section is based on Peter Haggett, *Geography: A Modern Synthesis*, third revised edition. New York: Harper and Row, 1983, ch. 25.

19 Hartshorne, *The Nature of Geography* (note 7), p. 147.

20 Of all the regional novelists, perhaps it is Thomas Hardy who has caught the imagination of geographers most vividly. Compare two papers written a generation apart, H. C. Darby, 'The regional geography of Thomas Hardy's Wessex'. *Geographical Journal*, Vol. 38 (1948), pp. 426–43; and B. P. Birch, 'Wessex, Hardy and the nature novelists'. *Transactions of the Institute of British Geographers*, Vol. 6 (1981), pp. 348–58.

21 John Fraser Hart, 'The highest form of the geographer's art'. *Association of American Geographers, Annals*, Vol. 72 (1982), pp. 1–29. Hart's paper led to a lively discussion and rejoinders which was reported in the *Annals*, Vol. 72 (1982), pp. 557–9 and Vol. 73 (1983), pp. 439–43.

22 D. R. Stoddart, 'To claim the high ground: geography for the end of the

century'. *Institute of British Geographers, Transactions*, New Series, Vol. 12 (1987), pp. 327–36.

23   Bowman, *Geography in Relation to the Social Sciences* (note 12), quotation from pp. 146–7.

24   Albrecht Penck, 'Geography amongst the earth's sciences'. *Proceedings of the American Philosophical Society*, Vol. 66 (1927), pp. 621–44, quotation from p. 633.

25   This section is drawn largely from Peter Haggett, *Locational Analysis in Human Geography*. London: Edward Arnold, 1966, pp. 2–4.

26   N. R. Hanson, *Patterns of Discovery: An Inquiry into the Conceptual Foundations of Science*. Cambridge: Cambridge University Press, 1958, p. 204.

27   M. Postan, 'The revulsion from thought'. *Cambridge Journal*, Vol. 1 (1948), pp. 395–408, quotation from p. 406.

28   Hanson, *Patterns of Discovery* (note 26), pp. 11–14.

29   P. B. Porter, 'Another puzzle picture'. *American Journal of Psychology*, Vol. 47 (1954), pp. 550–51, quotation from p. 551.

30   Charles Dickens, *A Christmas Carol in Prose. Being a Ghost Story of Christmas*. With illustrations by John Leech. London: Chapman and Hall, 1843.

CHAPTER 2: LEVELS OF RESOLUTION

1   Much of the effort of current theoretical physics is concerned with building unified theories between these weak and strong forces. String theories form one of the most intriguing bridges so far devised. See Stephen W. Hawking, 'The unification of physics'. In *A Brief History of Time: From the Big Bang to Black Holes*. London: Bantam, 1988, pp. 155–70.

2   P. Haggett, R. J. Chorley and D. R. Stoddart, 'Scale standards in geographical research: a new measure of areal magnitude'. *Nature*, Vol. 205 (1965), pp. 844–7.

3   This essay is partly based on a chapter by the author, 'Scale components in geographical problems', in R. J. Chorley and P. Haggett (eds), *Frontiers in Geographical Teaching: The Madingley Lecture for 1963*. London: Methuen, 1965, pp. 164–85. A valuable review of scale problems is given in David H. Miller, 'The factor of scale: ecosystem, landscape mosaic and region'. In K. A. Hammond, G. Macinko, and W. B. Fairchild (eds), *Sourcebook of the Environment: A Guide to the Literature*. Chicago: University of Chicago Press, 1978, pp. 63–88.

4   The six lectures were given in the Harkness Theatre of Columbia University, New York City, between 29 January and 6 February 1952. Sauer described them not as a 'well-polished abstract of accepted learning but a prospectus of that which is not securely within our grasp.' They were published as Carl O. Sauer, *Agricultural Origins and Dispersals* (The Bowman Memorial Lectures, Series Two). New York: American Geographical Society, 1952.

5   Edgar Anderson, *Plants, Man and Life*. London: Andrew Melrose, 1954, quotation, p. 120. Some of the delights of this book are reflected in the chapter titles: 'How to measure an avocado' is typical.

6   Ibid., p. 125.

7   The figure is based on the names of geographers recorded in E. Meynen, *Orbis Geographicus 1980/84: World Directory of Geography*. Wiesbaden: Franz Steiner, 1982.

8   R. Hartshorne, *Perspectives on the Nature of Geography*. London: John Murray, 1959, p. 21; Carl Ortwin Sauer, *Agricultural Origins and Dispersals*. New York: American Geographical Society, 1952.

9   J. C. Beaglehole, *The Life and Times of Captain James Cook*. London: The Hackluyt Society, 1974 (Volume IV of *The Journals of Captain James Cook on His Voyages of Discovery*). Map 5, facing p. 704.

10 Ibid., p. 713.

11 Ibid.

12 Joseph Conrad, 'Geography and some explorers'. In *Last Essays*, London: J. M. Dent, 1926, p. 17. Cited by David Stoddart, 'Geography, exploration and discovery'. In *On Geography and its History*, Oxford: Basil Blackwell, 1986, p. 143.

13 This problem is discussed at length in Haggett, 'Scale components' (note 3).

14 Peter Haggett, 'Land use and sediment yield in an old plantation tract of the Serra do Mar, Brazil'. *Geographical Journal*, Vol. 127, 1961, pp. 50–9.

15 Captain Richard Burton, *Explorations of the Highlands of Brazil with a Full Account of the Gold and Diamond Mines, Also Canoeing down 1500 Miles of the San Francisco from Savara to the Sea*. London: Tinsley Brothers, 1869, Vol. I, p. 112.

16 R. S. Platt, 'Field approach to regions'. *Association of American Geographers, Annals*, Vol. 25 (1935) pp. 153–74; R. S. Platt, *Latin America: Countrysides and United Regions*. New York: McGraw Hill, 1942.

17 R. M. Highsmith, O. H. Heintzelman, J. G. Jensen, R. D. Rudd and P. R. Tschirley, *Case Studies in World Geography*. Englewood Cliffs, N.J.: Prentice Hall, 1961.

18 One of the most useful reviews remains that by the geologist Bill Krumbein who had such an acute sense of the errors that could arise when working in a spatial framework. See, for example, the chapter on 'Geological sampling' in W. C. Krumbein and F. A. Graybill, *An Introduction to Statistical Models in Geology*. New York: McGraw-Hill, 1965, pp. 146–68. R. A. Fisher's classic early work was summarized in his *Statistical Methods for Research Workers*. Edinburgh: Oliver and Boyd, 1925. D. G. Kendall, the Cambridge statistician, has developed the idea of sampling as a set of 'stochastic traps' to identify and locate a spatial pattern. See D. G. Kendall, 'Foundations of a theory of random sets'. In E. F. Harding and D. G. Kendall, *Stochastic Geometry: A Tribute to the Memory of Rollo Davidson*. London: Wiley, 1970, pp. 323–30.

19 W. F. Wood, 'Use of stratified random samples in a land-use study'. *Association of American Geographers, Annals*, Vol. 45 (1955), pp. 350–67.

20 B. J. L. Berry, 'Sampling coding and storing flood-plain data'. *United States*

Department of Agriculture, Farm Economics Division, Agricultural Handbook, No. 237 (1962), pp. 1–27.

21 See for example, P. Haggett, 'Regional and local components in land-use sampling'. *Erdkunde*, Vol. 17 (1963), pp. 108–14; P. Greig-Smith, *Quantitative Plant Ecology*, 2nd edn. London: Butterworth, 1964.

22 P. J. Cook-Mozaffari, F. L. Ashwood, T. Vincent, D. Forman and M. Alderson, *Cancer Incidence and Mortality in the Vicinity of Nuclear Installations, England and Wales, 1959–80*. London: Her Majesty's Stationery Office (Office of Population Censuses and Surveys, Studies on Medical and Population Subjects, 51).

23 Standardized mortality ratios are a way of correcting crude death rates for the age structure of a population for each sex. Although simple in concept, their application is made difficult by the need to choose a standard population against which local death rates can be compared. See B. Benjamin, *Health and Vital Statistics*. London: George Allen and Unwin, 1968, pp. 92–9.

24 S. Openshaw, A. W. Craft, M. Charlton and J. M. Birch, 'Investigation of leukaemia clusters by use of a geographical analysis machine'. *Lancet*, 1 (1988), pp. 272–3.

25 G. E. P. Box, 'The exploration and exploitation of response surfaces'. *Biometrics*, Vol. 10 (1954), pp. 16–30.

26 L. L. Nettleton, 'Regions, residuals and structures'. *Geophysics*, Vol. 19 (1954), pp. 1–22; quotation from p. 10.

27 C. H. G. Oldham and D. B. Sutherland, 'Orthogonal polynomials: their use in estimating the regional effect'. *Journal of Geology*, Vol. 62 (1954), pp. 26–49; W. C. Krumbein, 'Trend-surface analysis of contour-type maps with irregular control-point spacing'. *Journal of Geophysical Research*, Vol. 64 (1959), pp. 823–34; R. J. Chorley and P. Haggett, 'Trend surface mapping in geographical research'. *Institute of British Geographers, Transactions*, Vol. 37 (1965), pp. 47–67.

28 J. S. Olsen and P. E. Potter, 'Variance components of cross-bedding direction in some basal Pennsylvanian sandstones of the Eastern Interior Basis: statistical methods'. *Journal of Geology*, Vol. 62 (1954), pp. 26–49.

29 R. J. Chorley, D. R. Stoddart, P. Haggett and H. O. Slaymaker, 'Regional and local components in the areal distribution of surface sand facies in the Breckland, eastern England'. *Journal of Sedimentary Petrology*, Vol. 36 (1966), pp. 209–20.

30 O. Duncan, R. P. Cuzzort and B. Duncan, *Statistical Geography: Problems in Analyzing Areal Data*. Glencoe, Ill.: Free Press, 1961. The statistical role of geography was emphasized by Dudley Stamp in his paper, 'Geographical agenda: a review of some tasks awaiting geographical attention', *Transactions, Institute of British Geographers*, Vol. 23 (1957), pp. 1–17; viz. 'The nearest parallel to the role of the geographer seems to me to be presented by that of the statistician' (quotation p. 1).

31 M. G. Kendall, 'The geographical distribution of crop productivity in England'. *Journal of the Royal Statistical Society*, Vol. 102 (1939), pp. 21–48.

32 M. Church, 'Records of recent geomorphological events'. In R. A. Cullingford,

D. A. Davidson and J. Lewin (eds), *Timescales in Geomorphology*. Chichester: John Wiley, 1980, pp. 14–29.

33 P. Haggett, 'Leads and lags in inter-regional systems: a study of cyclic fluctuations in the South West economy'. In M. D. I. Chisholm and G. Manners (eds), *Spatial Policy Problems of the British Economy*. Cambridge University Press, 1971, pp. 69–95; L. W. Hepple, 'Spectral techniques and the study of interregional economic cycles'. In R. F. Peel, M. D. I. Chisholm and P. Haggett (eds), *Processes in Physical and Human Geography: Bristol Essays*. London: Heinemann, 1975, pp. 392–408.

34 Peter Hall, 'The geography of the Fifth Kondratieff'. In Peter Hall and Ann Markusen (eds), *Silicon Landscapes*. Boston: Allen and Unwin, 1985, pp. 1–19. The original idea of the Kondratieff wave is given in N. D. Kondratieff, 'The long waves in economic life'. *Review of Economic Statistics*, Vol. 17 (1935), pp. 105–15.

35 H. H. McCarty, J. C. Hook and D. S. Knos, 'The measurement of association in industrial geography'. *University of Iowa, Department of Geography, Report No. 1* (1956), pp. 1–143.

36 B. B. Mandelbrot, 'How long is the coast of Britain? Statistical self-similarity and fractional dimension'. *Science*, Vol. 156 (1967), pp. 636–8. Mandelbrot has subsequently expanded his ideas in a series of books and papers, see for example his *Fractals: Form, Chance and Dimension*. San Francisco: Freeman, 1977; idem, *Fractal Geometry of Nature*. San Francisco: Freeman, 1982. The geographical implications of the work have been developed by Michael Batty, 'Randomness and recursion: fractal landscapes', in *Microcomputer Graphics*. London: Chapman and Hall, 1987, pp. 144–97. See also Michael F. Goodchild and David M. Mark, 'The fractal nature of geographic phenomena'. *Association of American Geographers, Annals*, Vol. 77 (1987), pp. 265–78. For an environmental application see W. E. H. Culling, 'Highly erratic spatial variability of soil-pH on Iping Common, West Sussex'. *Catena*, Vol. 13 (1986), pp. 81–98.

37 Hawking, 'The unification of physics' (note 1), pp. 155–70.

CHAPTER 3: THE ART OF THE MAPPABLE

1 The floods are described in J. A. Steers, 'The East Coast floods, January 31–February 1 1953'. *Geographical Journal*, Vol. 119 (1953), pp. 280–95; discussion pp. 295–8. The subsequent coastal changes resulting from the floods are analysed in J. A. Steers and A. T. Grove, 'Shoreline changes on the marshland coast of north Norfolk, 1951–53'. *Transactions of the Norfolk and Norwich Naturalists' Society*, Vol. 17 (1954), pp. 322–6. Both Spate and Peel were undergraduates at Cambridge in the early 1930s: O. H. K. Spate went on to be Professor of Geography in the Australian National University and the author of major works on India and the Pacific; R. F. Peel to be Professor at both Leeds and Bristol and an expert on arid-zone geomorphology. Of their successors at Scolt in the 1950s, Derek Brearley moved into educational

administration while M. D. I. Chisholm and R. J. Small became Professors of Geography at Cambridge and Southampton.

2 R. D. Hayes, 'A peasant economy in north-west Portugal'. *Geographical Journal*, Vol. 122 (1956), pp. 54–70. The members of the survey party in addition to Hayes and Haggett were P. A. Colinvaux, J. B. Nielson, A. Spivey and N. J. Spearing: see chapter 8 of this volume.

3 S. W. Wooldridge, *The Geographer as Scientist*. London: Nelson, 1956, p. 73. Cited by David Stoddart, *On Geography and its History*. Oxford: Basil Blackwell, 1986, p. 144.

4 The maps are described in P. Haggett, 'The edges of space'. In R. J. Bennett (ed.), *European Progress in Spatial Analysis*. London: Pion, 1981, pp. 51–70.

5 Multi-dimensional scaling as a mapping method is described in D. G. Kendall, 'The recovery of structure from fragmentary information', *Philosophical Transactions of the Royal Society of London, Series A*, Vol. 279 (1975), pp. 547–82. A review of geographical applications is given in A. C. Gatrell, 'Multidimensional scaling', in N. Wrigley and R. J. Bennett (eds), *Quantitative Geography*. London: Routledge and Kegan Paul, 1981, pp. 151–63 and in R. G. Golledge, *Multidimensional Scaling: Review and Geographical Applications*. Washington, D.C.: Association of American Geographers, 1972. (Commission on College Geography, TP-10).

6 For a discussion of conventional map projections and their properties, see Derek H. Maling, *Coordinate Systems and Map Projections*. New York: Marcel Dekker, 1979. Subsequent developments in cartography are reviewed in M. S. Monmonier, *Technological Transition in Cartography*. Madison: University of Wisconsin Press, 1985.

7 D. G. Kendall, 'Construction of maps from "odd bits of information"'. *Nature*, Vol. 231 (1971), pp. 158–9.

8 The document is a sixteenth-century copy of a thirteenth-century document. It is cited in full in Patricia Galloway, 'Restoring the maps of medieval Trondheim: a computer-aided investigation of the nightwatchmen's itinerary'. *Journal of Archaeological Science*, Vol. 5 (1978), pp. 153–65. I am grateful to Michael Blakemore of Durham University for drawing this example to my attention.

9 Ibid.

10 Peter Gould, 'Concerning a geographic education'. In David A. Lanegran and Risa Palm, *An Invitation to Geography*, 2nd edn. New York: McGraw Hill Book Company, 1978, pp. 202–25.

11 Werner Heisenberg, *Physics and Beyond: Encounters and Conversations*. New York: Harper and Row, 1972, p. 51. Cited by Yi-Fu Tuan, *Space and Place*. Minneapolis: University of Minnesota Press, 177; quotation from p. 4. See the general discussion of these terms in Peter Haggett, *Geography: A Modern Synthesis*, 3rd edn. New York: Harper and Row, 1983, p. 5.

12 J. L. Holloway, 'Smoothing and filtering of time series and space fields'. *Advances in Geophysics*, Vol. 4 (1958), pp. 351–89.

13 Regular grids have been widely used as a way of overcoming the problem of irregular collecting areas. One of the earliest examples is F. H. Perring and

S. M. Walters, *Atlas of the British Flora*. London: Botanical Society of the British Isles, 1962. Since 1962 the 10-km square of the National Grid has been used for other mapping schemes; see for example A. C. Jermy (ed.), *Atlas of Ferns of the British Isles*. London: Botanical Society of the British Isles, 1978, and J. T. R. Sharrock, *The Atlas of Breeding Birds in Britain and Ireland*. London: British Trust for Ornithology, 1976. There are some advantages in using regular hexagons rather than squares; see A. H. Robinson, J. B. Lindbergh and L. W. Brinkman, 'A correlation and regression analysis applied to total farm population densities in the Great Plains'. *Association of American Geographers, Annals*, Vol. 51 (1961), pp. 211–21. The problem of using non-grid matrices is illustrated in P. H. Rees, *Migration and Settlement: the United Kingdom*. Laxenburg, Austria: International Institute for Applied Systems Analysis, 1979.

14   A. D. Cliff, P. Haggett, J. K. Ord and G. R. Versey, *Spatial Diffusion: An Historical Geography of Epidemics in an Island Community*. Cambridge: Cambridge University Press, 1981, pp. 56–9.

15   S. Openshaw and P. J. Taylor, 'A million or so correlation coefficients: three experiments on the modifiable unit area problem'. In N. Wrigley (ed.), *Statistical Applications in the Spatial Sciences*. London: Pion, 1972, pp. 127–44.

16   Leslie Curry's original paper was 'A spatial analysis of gravity flows'. *Regional Studies*, Vol. 6 (1972), pp. 131–47. For an empirical test see D. A. Griffith and K. G. Jones, 'Explorations into the relationship between spatial structure and spatial interaction'. *Environment and Planning A*, Vol. 12 (1980), pp. 187–201.

17   W. C. Krumbein, 'A comparison of polynomial and Fourier models in map analysis'. Office of Naval Research, Geography Branch, ONR Task. No. 388-078, Contract 1228(36), TR-2. Evanston, Illinois: Northwestern University, 1966.

18   Krumbein, ibid., see also W. R. Tobler, 'Smooth pycnophylactic interpolations for geographical regions' (With comments and a rejoinder). *Journal of the American Statistical Association*, Vol. 74 (1979), pp. 519–36.

19   D. J. Unwin and N. Wrigley, 'Control point distribution in trend surface modelling revisited: an application of the concept of leverage'. *Institute of British Geographers, Transactions*, New Series, Vol. 12 (1987), pp. 147–60.

20   For a general review of the edge-effects problem see P. Haggett, 'Boundary problems in quantitative geography'. In Haruko Kishimoto and Walter Kummerly (eds), *Geographie und ihre Grenzen: eine Gedenkschrift zu Ehren von Hans Boesch*. Berne: Kummerly and Frey, 1980, pp. 59–68.

21   Haggett, 'The edges of space' (note 4), p. 60.

22   M. Choynowski, 'Maps based upon probabilities'. *Journal of the American Statistical Association*, Vol. 54 (1959), pp. 385–8.

23   B. Efron and C. Morris, 'Stern's paradox in statistics'. *Scientific American*, Vol. 236, No. 5 (1977), pp. 119–27. A more formal presentation is given in B. Efron, 'Biased versus unbiased estimations'. *Advances in Mathematics*, Vol. 16 (1975), pp. 259–77.

24 The best account is given in A. D. Cliff and J. K. Ord, *Spatial Processes: Models and Applications*. London: Pion. See also D. A. Griffith, 'Towards a theory of spatial statistics'. *Geographical Analysis*, Vol. 12 (1980), pp. 325–39 and R. R. Haining, 'Spatial autocorrelation problems'. In D. T. Herbert and R. J. Johnston (eds), *Geography and the Urban Environment*. Chichester: John Wiley, 1980, pp. 1–44.

CHAPTER 4: REGIONAL SYNTHESIS

1 Torsten Hägerstrand, *Innovation Diffusion as a Spatial Process*. Chicago: Chicago University Press, 1967; Donald W. Meinig, *On the Margins of the Good Earth: the South Australian Wheat Frontier, 1869–1884*. Chicago: Rand McNally, 1962; Walter Christaller, *The Central Places of Southern Germany*. Englewood Cliffs, N.J.: Prentice Hall, 1966; S. W. Wooldridge and D. L. Linton, *Structure, Surface and Drainage in South East England*. London: George Philip, 1939. The Hägerstrand and Christaller volumes are translations of works published earlier in Swedish and German respectively.

2 August Lösch, *The Economics of Location*, translated by W. H. Woglom. New Haven: Yale University Press, 1954.

3 Ibid., the quotation is from p. 3 of the preface to the first German language edition, written at Heidenheim (Wurttemberg) in the autumn of 1939.

4 Carl Sauer, 'The education of a geographer'. *Association of American Geographers, Annals*, Vol. 46 (1956), pp. 287–99; quotation, p. 289. Cited in the preface to Peter Haggett, *Locational Analysis in Human Geography*. London: Edward Arnold, 1965, p. vi.

5 Crematoria must be among the most unlikely places from which to start a research trail. One November morning I found myself at Canford cemetery in a leafy north-Bristol suburb attending the funeral service of a friend, a distinguished economist. As ever it was both a family and a university occasion and after the service, when the family cars had departed, I walked down the drive with Howard Bracey, an ex-geographer who had long since retired from the Agricultural Economics Unit at Bristol. Then in his seventies, he confessed that he was spending the rest of the day clearing out his garage and making a bonfire of all his old university and research papers.

I inquired anxiously whether these included his unique village services surveys conducted in the parishes of the counties of Gloucestershire, Somerset and Wiltshire. To my horror he replied that two counties had already gone up in smoke, and that Somerset was to follow them that afternoon 'if the wind was in the right direction'. Two days later our departmental van had collected three filing cabinets of old questionnaires, survey forms and correspondence and by the end of that year the Social Sciences Research Council had awarded a grant to put these archives in order and to carry out a parallel and comparative survey 'thirty years on'. Comparative surveys of this kind are rare and the full results of the surveys by Liz Mills are still being processed. See E. A. Mills, Changes in the spatial economy of an English county (Somerset, 1950–80). Doctoral dissertation. University of Bristol, 1989.

6 B. H. Farmer, *Pioneer Peasant Colonization in Ceylon*. Oxford: Oxford University Press, 1957; Charles A. Fisher, *South-east Asia: A Social, Economic and Political Geography*. London: Methuen, 1964. The experience of British geographers in World War II is reviewed in W. G. V. Balchin, 'United Kingdom geographers in the Second World War'. *Geographical Journal*, Vol. 153 (1987), pp. 159–80. The American experience is reviewed in Joseph A. Russell, 'Military geography'. In Preston E. James, Clarence F. Jones and John K. Wright (eds), *American Geography: Inventory and Prospect*. Syracuse: Syracuse University Press, 1954, pp. 484–97.

7 Much of the Icelandic work over the last decade is sumnmarized in three volumes: A. D. Cliff, P. Haggett, J. K. Ord and G. R. Versey, *Spatial Diffusion: An Historical Geography of Epidemics in an Island Community*. Cambridge: Cambridge University Press, 1981; A. D. Cliff, P. Haggett and J. K. Ord, *Spatial Aspects of Influenza Epidemics*. London: Pion, 1986 (see pp. 133–259); and A. D. Cliff and P. Haggett, *Atlas of Disease Distributions*. Oxford: Basil Blackwell, 1988 (see especially pp. 245–57, 268–73).

8 Cliff, Haggett, Ord and Versey, *Spatial Diffusion*, p. 52.

9 One of the clearest introductions remains Owen L. Davies, *Design and Analysis of Industrial Experiments*. London: Oliver and Boyd for Imperial Chemical Industries, 1956.

10 A. D. Cliff, P. Haggett and R. Graham, 'Reconstruction of diffusion processes at different geographical scales: the 1904 measles epidemic in northwest Iceland'. *Journal of Historical Geography*, Vol. 9 (1983), pp. 29–46.

11 The research was published in a series of *Occasional Papers* by the Department of Geography, University of Tasmania, between 1978 and 1979 under the general title of 'The Tasman Bridge collapse and its effects on metropolitan Hobart': see T. R. Lee and L. J. Wood on 'Methodology'; Lee on 'Social disruption and adjustments'; N. D. McGlashan on 'Medical facilities'; Wood on 'Shopping patterns' and Wood and Lee on 'Disruption of journey-to-work patterns'.

12 This section has its roots in a public lecture given in Christchurch, New Zealand, on 30 March 1979, while I was following in Andrew Clark's and Michael Wise's shoes as an Erskine Visiting Fellow at the University of Cantebury. It was subsequently published as 'Emerging trends in regional geography: a view from the touch line', in Philip Forer (ed.), *Futures in Human Geography*. Christchurch: New Zealand Geographical Society, 1980, pp. 1–11. I gratefully acknowledge the encouragement given by Barry Johnston and his colleagues at Christchurch to my chasing these elusive hares.

13 T. J. Chandler, *The Climate of London*. London: Hutchinson, 1965.

14 R. S. Platt, 'Field approach to regions'. *Association of American Geographers, Annals*, Vol. 25 (1935), pp. 153–74.

15 Glenn T. Trewartha, *The Earth's Problem Climates*, 2nd edn. Madison: University of Wisconsin Press, 1981.

16 P. Haggett, 'Towards a statistical definition of ecological range: the case of

Quercus suber'. Ecology, Vol. 45 (1964), pp. 622–5.

17 Michael Yakovlevich Nuttonson, *Ecological crop geography of the Ukraine and Ukrainian agro-climatic analogues in North America.* Washington, D.C.: American Institute of Crop Ecology, 1947. The Ukraine study was one of more than twenty major studies conducted by Nuttonson between 1947 and 1957.

18 The films referred to in this paragraph are *The Bridge on the River Kwai* (Great Britain, Columbia, 1957) and *Doctor Zhivago* (United States, Metro-Goldwyn-Mayer, 1965). The remake of *Beau Geste* I've been unable to trace. It would be wrong to think all films are shot in analogue locations; the Australian film industry has been meticulous in setting films in the exact location wherever possible. As an undergraduate I used to go twice a week to the movies and a fellow enthusiast in the year above me at my Cambridge college was the late Leslie Halliwell whose *Film Guide* (2nd edn, London: Granada, 1982) provides a note on over 10,000 films. I tried to persuade Halliwell to include the filming locations in future editions of his digest to assuage those geographers who want to know.

19 Peter Haggett, *Geography: A Modern Synthesis.* New York: Harper and Row, 1982, pp. 592–3.

20 Nevin M. Fenneman, 'Physiographic divisions of the United States'. *Association of American Geographers, Annals,* Vol. 6 (1916), pp. 19–98; J. F. Unstead, 'A system of regional geography'. *Geography,* Vol. 18 (1933), pp. 175–87.

21 David L. Linton, 'The delimitation of morphological regions'. *Institute of British Geographers, Transactions,* No. 14 (1949), pp. 86–7.

22 Derwent Whittlesey, 'Southern Rhodesia: An African compage'. *Association of American Geographers, Annals,* Vol. 46 (1956), pp. 1–97; idem, 'The regional concept and the regional method'. In Preston E. James, Clarence F. Jones and John K. Wright (eds), *American Geography: Inventory and Prospect.* Syracuse: Syracuse University Press, 1954, pp. 19–69.

23 Whittlesey, *Southern Rhodesia* (note 22).

24 J. H. Bird, 'Scale in regional study; illustrated by brief comparisons between the western peninsulas of England and France'. *Geography,* Vol. 41 (1956), pp. 25–38.

25 A. K. Philbrick, 'Principles of areal functional organization in regional human geography'. *Economic Geography,* Vol. 33 (1957), pp. 299–336.

26 J. H. Paterson, 'Writing regional geography'. *Progress in Geography,* Vol. 6 (1974), pp. 1–26.

27 William R. Mead's writings include *Farming in Finland.* London: Athlone Press, 1953; and *An Historical Geography of Scandinavia.* London: Academic Press, 1981. James J. Parsons, *Antioqueno Colonization in Western Colombia,* revised edition. Berkeley: University of California Press, 1968. H. C. Brookfield with Doreen Hart, *Melanesia: A Geographical Interpretation of an Island World.* London: Methuen, 1971.

28 O. H. K. Spate, *India and Pakistan: A General and Regional Geography.* London: Methuen, 1954. The third edition (1967) was revised with A. T. A. Learmonth as co-author.

29 Ibid., p. 407.

30 Ibid., p. 409.

31 Charles Fisher, 'Whither regional geography?'. *Geography*, Vol. 55 (1970), pp. 373–89; quotation from p. 376.

32 Walter Prescott Webb, *The Great Plains*. London: Oxford University Press, 1931; Donald W. Meinig, *Imperial Texas: An Interpretive Essay in Cultural Geography*. Austin: University of Texas Press, 1969.

33 H. C. Darby, 'The problem of geographical description'. *Institute of British Geographers, Transactions*, Vol. 30 (1962), pp. 1–14; quotations from p. 2.

34 Douglas Wilson Johnson, *Battlefields of the World War. Western and Southern Fronts. A Study in Military Geography*. New York: Oxford University Press, 1921. (American Geographical Society, Research Series, No. 3, 648 pp.) Quotation from p. 1.

35 Yi-Fu Tuan, 'Geopiety: a theme in man's attachment to nature and to place'. In David Lowenthal and Martyn J. Bowden (eds), *Geographies of the Mind: Essays in Historical Geography in Honor of John Kirtland Wright*. New York: Oxford University Press, 1976, pp. 11–40.

36 Translated from a modern popular almanac, *Reimmichls Volkskalender of South Tyrol* by Leonard Dorb in his *Patrioticism and Nationalism: Their Psychological Foundations*. New Haven: Yale University Press, 1964, p. 194. Cited by Yi-Fu Tuan, *Space and Place*, Minnesota, 1977, p. 4.

37 Paul Tillich, *On the Boundary: An Autobiographical Sketch*. New York: Scribner, 1966; quotation from p. 17.

38 Paul Tillich, *Systematic Theology*, Volume III. Chicago: University of Chicago Press, 1963; quotation p. 318.

39 Typical of this hatred is the reaction of M. Butor, *Passing Time and a Change of Heart*. New York: Simon and Schuster, 1969; quotation, p. 48. 'From the very first I had felt this town to be unfriendly, unpleasant, a treacherous quicksand; ... I gradually felt its lymph seeping into my blood, its grip tightening, my present existence growing rudderless, amnesia creeping over me, that I began to harbour that passionate hatred towards it which, I am convinced, was in part a sign of my contamination by it.' See also Yi-Fu Tuan, *Landscapes of Fear*. New York: Pantheon, 1979.

40 John Muir, *My First Summer in the Sierra, with Illustrations Made by the Author in 1869 and from Photographs by H. W. Gleason*. London: Constable, 1911; quotation from p. 157.

41 D. Steiner, 'A multivariate statistical approach to climatic regionalization and classification'. *Tijdschrift van het Koninklijk Nederlandsch Aardrijkskrundig Genootschap Tweede Reeks*, Vol. 82 (1965), pp. 329–47. For the limitations of quantitative analysis in classification, a classic paper remains R. J. Johnston, 'Choice in classification: the subjectivity of objective methods'. *Association of American Geographers, Annals*, Vol. 58 (1968), pp. 575–89.

42 Jay W. Forrester, *World Dynamics*. Cambridge, Mass.: Wright-Allen Press, 1971.

43 Joseph E. Schwartzberg (ed.), *A Historical Atlas of South Asia*. Chicago: University of Chicago Press, 1978; J. S. Duncan (ed.), *Atlas of Victoria*. Melbourne: Victorian Government Printing Office, 1982.

CHAPTER 5: THE ARROWS OF SPACE

1   For the spread of diseases in the south Pacific see Norma McArthur, *Island Populations of the Pacific*. Canberra: Australian National University Press, 1967. The work in Fiji is described in A. D. Cliff and P. Haggett, *The Spread of Measles in Fiji and the Pacific: Spatial Components in the Transmission of Epidemic Waves through Island Communities*. Canberra: Australian National University (Research School of Pacific Studies, Department of Human Geography, Publication HG.18), 1985.

2   'Men, women and children, are today dying by thousands in Fiji. What are the Executive doing in the matter? Nothing! They are endeavouring to hide their own sluggish supineness by throwing the responsibilities on the Wesleyan Missionaries.' *Fiji Times*, Levuka, 3 April 1875.

3   For a fuller account, see Cliff and Haggett, *The Spread of Measles* (note 1), pp. 21–9.

4   Torsten Hägerstrand, 'On Monte Carlo simulation of diffusion'. *Northwestern University, Studies in Geography*, Vol. 13 (1967), pp. 1–32.

5   See for example P. Haggett, A. D. Cliff and A. E. Frey, *Locational Analysis in Human Geography*, 2nd edn. London: Edward Arnold, 1977, pp. 234–7.

6   R. L. Morrill, 'The negro ghetto: problems and alternatives'. *Geographical Review*, Vol. 55 (1965), pp. 339–61; M. Levison, G. R. Ward and J. W. Webb, *The Settlement of Polynesia: A Computer Simulation*. Minneapolis: University of Minnesota Press, 1973.

7   Ira S. Lowry, 'A short course in model design'. *Journal of the American Institute of Planners*, Vol. 30 (1965), pp. 22–53.

8   Rowland Tinline, A Simulation Study of the 1967–8 Foot-and-Mouth Epizootic in Great Britain. Doctoral dissertation, University of Bristol, England, 1972. See also the review by A. D. Cliff and P. Haggett, 'Spatial aspects of epidemic control'. *Progress in Human Geography*, Vol. 13 (1989), pp. 315–47.

9   A. D. Cliff, P. Haggett, J. K. Ord and G. R. Versey, *Spatial Diffusion: An Historical Geography of Epidemics in an Island Community*. Cambridge: Cambridge University Press, 1981, p. 1.

10 M. S. Bartlett, 'Measles periodicity and community size'. *Journal of the Royal Statistical Society*, Series A, Vol. 12 (1957), pp. 49–70.

11 F. L. Black, 'Measles periodicity in insular populations: critical community size and its evolutionary implications'. *Journal of Theoretical Biology*, Vol. 11 (1966), pp. 207–11.

12 Cliff et al. *Spatial Diffusion* (note 9).

13 All populations refer to 1970, a convenient mid-point within the time period. The present population of Reykjavik is around 110,000.

14 The models employed are discussed in Cliff et al., *Spatial Diffusion*, ch. 6. See also A. D. Cliff, P. Haggett and J. K. Ord, 'Forecasting epidemic pathways for measles in Iceland: the use of simultaneous equations and logit models', *Ecology of Disease*, Vol. 2 (1983), pp. 377–96.

15 P. Haggett, *Potential Applications of Spatial Forecasting Models to MMWR*

Data. Atlanta, Georgia: US Department of Health, Centers for Disease Control (CDC–CSCA Report), 1982.

16   Ibid., p. 32.

17   A. D. Cliff, P. Haggett and J. K. Ord, *Spatial Aspects of Influenza Epidemics*. London: Pion.

18   August Lösch, *The Economics of Location*. New Haven: Yale University Press, 1954; see p. 407.

19   Peter J. Smailes, The effects of changes in agriculture upon the service structure of South Australia country service towns, 1945–1974. University of South Australia, Department of Geography, unpublished report, p. 72. See also John B. Parr, 'A note on the size distribution of cities over time'. *Journal of Urban Economics*, Vol. 18 (1985), pp. 199–212.

20   For an overall summary of geographers' work on economic fluctuations, see R. L. Martin and N. A. Spence, 'Economic geography'. In N. J. Wrigley and R. J. Bennett (eds), *Quantitative Geography: A British View*. London: Routledge and Kegan Paul, 1981, pp. 332–5.

21   R. Fels and C. E. Hinshaw, *Forecasting and Recognizing Business Cycle Turning Points*. New York: National Bureau for Economic Research, 1968.

22   L. J. King, E. Casetti and D. Jeffrey, 'Economic impulses in a regional system of cities'. *Regional Studies*, Vol. 3 (1969), pp. 213–18.

23   Leslie Hepple, 'Spectral techniques and the study of interregional economic cycles'. In Ronald Peel, Michael Chisholm and Peter Haggett (eds), *Processes in Physical and Human Geography: Bristol Essays*. London: Heinemann, 1975, pp. 392–410.

CHAPTER 6: FAMILY HISTORY

1   The growth of geography in the college is described in J. A. Steers, 'Geography at St Catharine's College', *St Catharine's College Society Magazine* (1983), pp. 20–3.

2   This essay is based in part on chapter 25 of Peter Haggett, *Geography: A Modern Synthesis*, 3rd rev. edn. New York: Harper and Row, 1983.

3   I find one of the most readable accounts of this period remains H. F. Tozer, *A History of Ancient Geography*. Cambridge: Cambridge University Press, 1897. Our neglect of the earliest roots of geography is reflected by the fact that the last borrowing date on the library copy I checked was forty years ago.

4   John Ziman, *The Force of Knowledge: The Scientific Dimension of Society*. Cambridge: Cambridge University Press, 1976, pp. 120–45.

5   Peter Haggett, 'Torsten Hägerstrand'. In Alan Bullock and R. B. Woodings (eds), *The Fontana Biographical Dictionary of Modern Thought*. London: Collins, 1983, p. 298. See also Allan Pred, 'Postscript'. In Torsten Hägerstrand, *Innovation Diffusion as a Spatial Process*. Chicago: University of Chicago Press, 1967.

6   Torsten Hägerstrand, 'Diorama, path and project'. *Tijdschrift voor Economische en Sociale Geographie*, Vol. 73 (1982), pp. 323–39; quotation from p. 335.

See also Torsten Hägerstrand, 'Space, time and human conditions'. In A. Karlqvist, L. Lundqvist and F. Snickars (eds), *Dynamic Allocation of Urban Space*. Farnborough: Saxon House, 1975, pp. 3–14; quotation from p. 8.

7 Torsten Hägerstrand, 'In search for the sources concepts'. In Anne Buttimer (ed.), *The Practice of Geography*. London: Longman, 1983, pp. 238–356.

8 The richness of the biographical material on geographers has been greatly helped by the series edited by the late Walter Freeman for the IGU Commission on the History of Geographical Thought: T. W. Freeman (ed.), *Geographers: Bibliographical Studies*. London: Mansell, Vol. 1–, 1977. There are now twelve volumes with some 200 biographies.

9 The Travellers' Club had a reputation for being one of the most exclusive in London, blackballing both the imperialist Cecil Rhodes and the novelist William Thackeray. The verse is from a doggerel by Theodore Hooke; see P. H. Ditchfield, *London's West End*. London: Jonathan Cape, 1925; quotation, p. 66.

10 T. W. Freeman, 'The Royal Geographical Society and the development of geography'. In E. H. Brown (ed.), *Geography: Yesterday and Tomorrow*. Oxford: Oxford University Press, 1980, pp. 1–99.

11 J. K. Wright, *Geography in the Making: American Geographical Society, 1851–1951*. New York: AGS, 1952.

12 Preston E. James and Geoffrey J. Martin, *The Association of American Geographers: The First Seventy-Five Years, 1904–1979*. Washington D.C.: Association of American Geographers, 1978; R. Steel, *The Institute of British Geographers: The First Fifty Years*. London: IBG, 1985.

13 A good summary of the work of the National Geographic Society is given in the forewords to the occasional indexes; see *National Geographic Index 1888–1946* and *National Geographic Index 1947–1969*. Washington, D.C.: National Geographic Society, 1967 and 1970.

14 D. R. Stoddart, 'Growth and structure of geography'. *Institute of British Geographers, Transactions*, Vol. 41 (1967), pp. 1–20.

15 Ibid., p. 3.

16 The Chicago geographer, Chauncy D. Harris, has made massive contributions to our understanding of the geographical literature. See C. D. Harris and J. D. Fellmann, *International List of Geographical Serials*, 3rd edn. Chicago: University of Chicago, 1980 (University of Chicago, Department of Geography, Research Paper, No. 193).

17 A. C. Gatrell and Anthony Smith, 'Networks of relations among a set of geographical journals'. *The Professional Geographer*, Vol. 36 (1984), pp. 300–7.

18 H. D. White and B. C. Griffiths, 'Authors as markers of intellectual space: co-citation in studies of science, technology and society'. *Journal of Documentation*, Vol. 38 (1982), pp. 255–72.

19 The parallels between the spread of epidemics and of rumours is explored by Karl Dietz in 'Epidemics and rumours: a survey'. *Journal of the Royal Statistical Society*, Series A, Vol. 130 (1967), pp. 505–28.

20 D. Edge, 'Quantitative measures of communication in science: a critical

review', *History of Science*, Vol. 17 (1979), pp. 102–34.

21 The impact factor of a journal is given by the number of citations in year x for papers published in the two previous years (x–1, and x–2) divided by the number of source items published in the two previous years. Assume the hypothetical journal, *Journal of Lost Regions*, had 230 citations in *other* journals in 1990 from the papers it had published in 1988 and 1989; we might regard this as its 'export' of ideas. This would have to be divided by the number of source items appended to the papers it had itself published in 1988 and 1989; i.e. ideas 'imported'. If this latter number was, say 183, then the impact factor would be (230/183), or 1.26. It would be a net exporter, since its value was greater than one. Values of the impact factor in the text are rounded. See J. W. R. Whitehand, 'The impact of geographical journals: a look at the ISI data'. *Area*, Vol. 16 (1984), pp. 185–7.

22 A. D. Cliff and P. Haggett, 'Island epidemics'. *Scientific American*, Vol. 250, No. 5 (May 1984), pp. 138–47.

23 Neil Wrigley and Stephen Matthews, 'Citation classics and citation levels in geography'. *Area*, Vol. 18 (1986), pp. 185–94. See also J. W. R. Whitehand, 'Contributors to the recent developments and influence of human geography: what citation analysis suggests', *Transactions, Institute of British Geographers*, Vol. 10 (1985), pp. 22–34.

24 D. R. Stoddart, 'The RGS and the foundations of geography at Cambridge'. *Geographical Journal*, Vol. 141 (1975), pp. 216–39.

25 James and Martin, *The Association of American Geographers*, p. 6. (note 12)

26 Data on the distribution of geography departments in North America with staff lists are given in the annual volumes of the *Guide to Departments of Geography in the United States and Canada* (Washington D.C.: Association of American Geographers) and for Commonwealth countries in the annual *Commonwealth Universities Yearbook* (London: Association of Commonwealth Universities). For the rest of the world a useful source is Emil Meynen (ed.), *Orbis Geographicus 1980/84*. Wiesbaden: Franz Steiner Verlag, 1982.

27 Ronald Peel, 'The Department of Geography, University of Bristol, 1925–75'. In Ronald Peel, Michael Chisholm and Peter Haggett (eds), *Processes in Physical and Human Geography: Bristol Essays*. London: Heinemann, 1975, pp. 411–17.

28 *The Wisconsin Geographer*, Vol. 16 (1978), p. 13.

29 Berkeley Geography, 1988. Berkeley, California: Department of Geography. Pamphlet, p. 24.

30 The 'Admiralty Handbooks' were prepared by an interdisciplinary team during World War II by the Naval Intelligence Division of the British Admiralty. They contain a wealth of topographic detail. For example, the four volumes covering the Pacific Islands published by His Majesty's Stationery Office in 1945 consist of a General Survey, the Western Pacific and the Eastern Pacific (in two volumes). Altogether the Pacific volumes alone run to 2,600 pages.

CHAPTER 7: SHIFTING STYLES

1   Mrs Thornburgh is a character in Mrs Humphry Ward's novel *Robert Elsmere* (Book I, ch. 2) published in 1888. Mercifully, the phrase is the sole entry from the volume that has found its way into the Oxford Dictionary of Quotations.

2   Until its move to Madingley in 1960, Geophysics occupied some of the ground floor of the Department of Geography in Downing Place, Cambridge. The leading scientists in the Geophysics group were Professor Edward Bullard, Dr Keith Runcorn and Dr B. C. Browne.

3   The reversal of views on plate tectonics is described in B. Jones, 'Plate tectonics: a Kuhnian case?'. *New Scientist*, Vol. 63 (1974), pp. 536–8. See also Roy Haines-Young and James Petch, *Physical Geography: Its Nature and Methods*. London: Harper and Row, 1986, pp. 76–9.

4   This essay is based in part on two earlier papers, Peter Haggett, 'Mid-term futures for geography'. *Monash Publications in Geography*, No. 16 (1977), p. 25; and, idem, 'Geography in a steady-state environment'. *Geography*, Vol. 62 (1977), pp. 159–67.

5   John Ziman, 'Can scientific knowledge be an economic category?' *Minerva*, Vol. 12 (1974), pp. 384–8.

6   D. R. Stoddart, 'Growth and structure of geography'. *Institute of British Geographers, Transactions*, Vol. 41 (1967), pp. 1–20; Derek de Solla Price's original book, *Little Science, Big Science* was published in 1963; it has recently been republished posthumously together with some later essays as *Little Science, Big Science ... and Beyond*. New York: Columbia University Press, 1986.

7   Stoddart, 'Growth and structure of geography', p. 2. (note 6)

8   J. J. Parsons, 'Carl Ortwin Sauer, 1889–1975'. *Year Book of American Philosophical Society*, 1975, pp. 163–7. Since Sauer's death there has been a massive growth of interest in his work. Among the many papers see Michael Williams, '"The apple of my eye." Carl Sauer and historical geography'. *Journal of Historical Geography*, Vol. 9 (1983), pp. 1–28; Martin S. Kenzer, 'Milieu and the "intellectual landscape". Carl O. Sauer's intellectual heritage'. *Association of American Geographers, Annals*, Vol. 75 (1985), pp. 258–70.

9   So far, those who have worked with Andrew Cliff and myself include archivists, cartographers, computer scientists, epidemiologists, linguists and statisticians.

10 'Under the dual-support system, the university contributes the time of its academic staff, meets the cost of central overheads, covers the ordinary running costs and provides the equipment and materials that would reasonably be expected as standard in the "well-found" laboratory. The Research Councils pay the additional costs attributable to their projects.' Advisory Board for the Research Councils, *A Strategy for the Science Base*. London: Her Majesty's Stationery Office, 1987, p. 5.

11 Pentti Yli-Jokipii, 'Trends in Finnish Geography in 1920–1979 in the light of the journals of the period'. *Fennia*, Vol. 160 (1982), pp. 95–193. Elisabeth Lichtenberger, 'The German-speaking countries'. In R. J. Johnston and P.

Claval, *Geography since the Second World War: An International Survey*.
Beckenham: Croom Helm, 1984, pp. 158–84.

12   Sir George Thomson, *The Foreseeable Future*. Cambridge: Cambridge University Press, 1955.

13   John Ziman, Science and the Steady State. London, unpublished report.

14   Charles Fisher, 'Whither regional geography?'. *Geography*, Vol. 55 (1970), pp. 373–89; quotation from p. 376.

15   J. H. Paterson, 'Writing regional geography'. *Progress in Geography*, Vol. 6 (1974), pp. 1–26; quotation from p. 3.

16   J. H. Paterson, *North America: A Geography of Canada and the United States*, 7th edn. Oxford: Oxford University Press, 1984. It was first published in 1960.

17   A brief but helpful introduction to the problems of crossing the micro–meso–macro gaps is given in A. G. Wilson and M. J. Kirkby, *Mathematics for Geographers and Planners*, 2nd edn. Oxford: Clarendon Press, 1980. The difference between microstates and macrostates also lies at the heart of maximum entropy modelling, see A. G. Wilson, *Entropy in Urban and Regional Modelling*. London: Pion, 1970.

18   R. Hartshorne, *The Nature of Geography: A Critical Survey of Current Thought in the Light of the Past*. Lancaster, Pa.: Association of American Geographers, 1939.

19   David Harvey, *Explanation in Geography*. London: Edward Arnold, 1969. The book was based on a series of seminars given at Bristol University in the period when Harvey was a lecturer there. It is important to note that within only a few years Harvey's work had progressed along very different lines; see for example David Harvey, *Social Justice and the City*. London: Edward Arnold, 1973.

20   A good retrospective review of the period is provided in Bill Macmillan (ed.), *Remodelling Geography*. Oxford: Basil Blackwell, 1989.

21   Derek Gregory, *Ideology, Science and Human Geography*. London: Hutchinson, 1978.

22   Thomas S. Kuhn, *The Structure of Scientific Revolutions*. Chicago: Chicago University Press, 1962; and 2nd edn, 1970; see also a review of Kuhn's ideas in I. Latakos and A. Musgrave (eds), *Criticism and the Growth of Knowledge*. Cambridge: Cambridge University Press, 1970.

23   P. Haggett and R. J. Chorley, 'Models, paradigms and the new geography'. In R. J. Chorley and P. Haggett (eds), *Models in Geography: The Second Madingley Lectures*. London: Methuen, 1967, pp. 20–41.

24   Peter Haggett, 'Spatial forecasting: a view from the touchline'. In R. L. Martin, N. J. Thrift and R. J. Bennett (eds), *Towards the Dynamic Analysis of Spatial Systems*. London: Pion, 1978, pp. 205–10.

25   René Thom, *Structural Stability and Morphogenesis*. Reading, Mass.: Benjamin, 1975.

26   Ilya Prigogine and Isabelle Stengers, 'Order through fluctuations' in *Order Out of Chaos: Man's New Dialogue with Nature*. London: Heinemann, 1984, pp. 177–212. The geographical applications are described in P. M. Allen and M. Sanglier, 'Dynamic model of urban growth'. *Journal for Social and Biological*

Structures, Vol. 1 (1978), pp. 265–80; idem, 'Urban evolution, self-organization and decision-making'. *Environment and Planning A*, Vol. 13 (1981), pp. 167–83; and Denise Pumain, Léna Sanders and Thérèse Saint-Julien, *Villes et Auto-Organization*. Paris: Economica, 1989. See also the account in John P. Briggs and F. David Peat, *Looking Glass Universe*. London: Fontana, 1984.

27   Peter Haggett, 'Revolutions and quantitative geography'. Address to the Sixth European Colloquium on Theoretical and Quantitative Geography, Chantilly, France, 6 September 1989, p. 12.

28   The literature on chaos theory with its broad-ranging sweep from oscillating pendulums through epidemics to snowflake structures is likely to have a significant impact on geographic modelling in the 1990s. A useful survey of the mathematical ideas is provided by H. Bruce Stewart and J. M. Thompson, *Nonlinear Dynamics and Chaos*. Chichester: Wiley, 1986. A lively account by New York Times science reporter is found in James Gleick, *Chaos: Making a New Science*. New York: Cardinal, 1987.

29   Prigogine and Stengers, 'Order through fluctuations' (note 26), p. 208.

30   O. H. K. Spate, 'Palaeoclimates of geographical thought'. *Australian Geographer*, Vol. 14 (1978), pp. 1–7. Spate describes the paper engagingly as a 'slight essay in nostalgia' on the offprint sent to the author.

CHAPTER 8: GEOGRAPHY FUTURE

1   The Centenary Celebration was held at Cambridge on Saturday 2 July 1988. The hundred-year period was measured from 18 February 1888 when General Sir Richard Strachey, the president of the Royal Geographical Society, gave the first of four lectures on 'Principles of geography'. This and the bizarre (and sometimes hilarious) events surrounding the establishment of geography at Cambridge are beautifully told by D. R. Stoddart, 'The foundations of geography at Cambridge'. In *On Geography and its History*. Oxford: Blackwell, 1986, pp. 77–127. By an odd coincidence Sir Richard Strachey is buried in the Strachey family chapel of my home parish church at Chew Magna in the county of Somerset (the ersatz county of Avon since 1974). The large brass plaque on the north wall of the chapel recalls the achievements of his life, describing him as 'a scientific geographer'. I've yet to find another church memorial to a geographer which uses exactly that phrase.

2   Montesquieu, *Lettres persanes* (1721). Cited by Betrand de Jouvenel, *The Art of Conjecture*. London: Weidenfeld and Nicholson, 1967; quotation from p. 13. Peter Haggett, 'Forecasting alternative spatial, ecological and regional futures: possibilities and limitations'. In Richard J. Chorley (ed.), *Directions in Geography*. London: Methuen, 1972, pp. 217–36.

3   B. J. L. Berry, 'The geography of the United States in the year 2000'. *Institute of British Geographers, Transactions*, No. 51 (1970), pp. 22–53.

4   Samuel Johnson, according to James Boswell's *Life of Johnson*, Vol. I, p. 463 (31 July 1763). Johnson's comment referred to women preaching.

5 World Meteorological Organization, 'The present status of long-range forecasting in the world'. *World Meteorological Organization, Technical Note*, No. 48, 1970.

6 University Grants Committee, *Annual Survey 1985–86*. London: Her Majesty's Stationery Office, 1987. Annual reports or surveys have been published from 1919 up to 1979 under a variety of titles. Some of the background reports which set the context for these changes are: University Grants Committee, *A Strategy for Higher Education into the 1990s*. London: Her Majesty's Stationery Office, 1984; Committee of Vice-Chancellors and Principals (CVCP), *Report of the Steering Committee for Efficiency Studies in Universities* (the 'Jarratt Report'). London: CVCP, 1985. See also a report by a geographer, Keith Clayton, *The Measurement of Research Expenditure in Higher Education*. London: Department of Education and Science, 1987.

7 *Review of the University Grants Committee*. Report of a committee under the chairmanship of Lord Croham. London: Her Majesty's Stationery Office, 1987; quotation, p. 20.

8 The UGC funding model which controls the Government block grant to British universities has been described in a series of public letters from the chairman to vice-chancellors and principals. In essence, the funding is broken into two components, teaching (T) and research (R), and for each of the 36 subject cost centres the student load is multiplied by a unit of resource. The units of resource have not been published but reflect the cost of each subject, with clinical medicine and veterinary studies representing one end of the cost spectrum, and law and some humanities the other. The research component of the Government grant is more complex and its importance varies from subject to subject. It is made up of three components, a basic staff-related element (SR), a supplemental element which is geared to the success of the university department in raising outside research funding (CR and DR), and a small amount related to the quality of the research as judged by a UGC subject panel (JR).

9 Among the major reports are E. R. Oxburgh, P. Haggett, D. A. L. Jenkins, Sir Alan Muir Wood and Sir Frederick Stewart, *Strengthening University Earth Sciences*. London: University Grants Committee, 1987; Sir Peter Parker, *'Speaking for the Future': A Review of the Requirements of Diplomacy and Commerce for Asian and African Languages and Area Studies*. London: University Grants Committee, 1986.

10 Advisory Board for the Research Councils, *A Strategy for the Science Base*. London: Her Majesty's Stationery Office, 1987.

11 There is considerable evidence that this is already happening. Over the last four years the external funding for British geography departments has roughly doubled.

12 I elected to give my inaugural address at Bristol in 1967 on the impact computers were then having on geographical research. A shortened version of this was later published as 'Geographical research in a computer environment'. *Geographical Journal*, Vol. 135 (1969), pp. 500–9.

13 M. Levison, G. R. Ward and J. W. Webb, *The Settlement of Polynesia: A Computer Simulation*. Minneapolis: University of Minnesota Press, 1973.

14   S. Openshaw, A. W. Craft, M. Charlton and J. M. Birch, 'Investigation of leukaemia clusters by use of a geographical analysis machine'. *Lancet*, 1 (1988), pp. 272–3.

15   E. C. Barrett and L. F. Curtis, *Introduction to Environmental Remote Sensing*, 2nd edn. London: Chapman and Hall, 1982; see pp. 307–22.

16   This view is most forcefully expressed by Michael Eliot Hurst, 'Geography, social science and society: towards a de-definition'. *Australian Geographical Studies*, Vol. 18 (1980), pp. 3–21; idem, 'Geography has neither existence nor future'. In R. J. Johnston (ed), *The Future of Geography*. London: Methuen, 1985. See also Peter Haggett, 'Postscript to an antipodean reaction'. *Monash University, Department of Geography, Occasional Papers*, No. 17 (1977), pp. 31–2.

17   Carl O. Sauer, 'The education of a geographer'. *Association of American Geographers, Annals*, Vol. 46 (1956), pp. 287–99; quotation from p. 295.

18   Madingley Hall, an Elizabethan country house near Cambridge, was the scene during the 1960s of a number of summer schools for school teachers on changes in geography. The course was organized by Ray Pahl, now Professor of Sociology in the University of Kent, for the Cambridge University Extra-Mural Board. The summer courses for 1963 were written up as Richard J. Chorley and Peter Haggett (eds), *Frontiers in Geographical Teaching*. London: Methuen, 1965. Subsequent courses led to a second volume, idem, *Models in Geography: The Second Madingley Lectures*. London: Methuen, 1967. Madingley was also eventually to lead to one hardback series (*Progress in Geography*, 1969–76) which split into two parallel journals (*Progress in Physical Geography* and *Progress in Human Geography*, Vol. 1–, 1977).

19   John Galsworthy: 'History tells us that the status quo is of all things the most liable to depart; the millennium of all things the least likely to arrive'. Cited by Isaiah Bowman, *Geography in Relation to the Social Sciences*. New York: Charles Scribner's Sons, 1934, p. 6.

20   For a recent review of the impact of the computer on mapping, see Mark S. Monmonier, *Technological Transition in Cartography*. Minneapolis: University of Minnesota Press, 1985.

21   The way in which climatic research at all spatial scales is now affecting socioeconomic assessments is shown in Robert W. Kates (ed.), *Climate Impact Assessment: Studies of the Interactions of Climate and Society*. Chichester: John Wiley, 1985.

22   Some of the attempts to bring a regional focus to modelling have been pioneered by researchers outside geography. The ways in which biologists and economists are now co-operating is reflected in Leon C. Braat and Wal F. J. van Lierop, *Economic-Ecologic Modeling*. Amsterdam: North Holland, 1987 (Studies in Regional Science and Urban Economics, Vol. 16).

23   For a recent review of this area, see Dietrich Denecke and Gareth Shaw (eds), *Urban Historical Geography: Recent Progress in Britain and Germany*. Cambridge: Cambridge University Press, 1988.

24   See for example Bruce Mitchell and Dianne L. Draper, *Relevance and Ethics in Geography*. London: Longman, 1982.

25 There is now a flood of literature from the environmental remote-sensing community on the global monitoring of land-use change. See for example D. T. Lindgren, *Land Use Planning and Remote Sensing*. Dordrecht: Martinus Nijhoff, 1984; J. Denegre (ed.), *Thematic Mapping from Satellite Imagery*. London: International Cartographic Association, 1988. It is worth noting that a start had been made much earlier by the geographer, Dudley Stamp, in his initiation of the World Land Use Survey in 1951. See Christopher Board, 'Land use surveys: principles and practice', in *Land Use and Resources: A Memorial to Sir Dudley Stamp*. London: Institute of British Geographers, 1968, pp. 29–41.

26 See for example A. D. Cliff and J. K. Ord, 'The comparison of means when samples consist of spatially autocorrelated observations'. *Environment and Planning, A*, Vol. 7 (1975), pp. 725–34; idem, 'Autocorrelation and inferential statistics', in *Spatial Processes: Models and Applications*. London: Pion, 1981.

27 'Whenever you get into a back area I would be very glad to have photographs of either field samples or crib samples of maize … I am getting increasingly excited about the possibilities of a very simple kind of evidence shedding light on the origins of agriculture. I have for some years taken a keen interest in the kind of plants from gardens around the houses of ordinary, everyday people … These vague suspicions would become real data if we could get rough drawings to scale of the kind of things which are being grown around village or country houses with examples of typical ones. I now have pretty good records from one place or another from Mexico, Honduras, Guatemala, Costa Rica, and Colombia, and, of course, Ethiopia. If you could make some kind of record, be it ever so amateurish, just as long as it was detailed and gave some common, everyday description of the plants and what they were used for it would certainly break new ground.' Letter dated 14 May 1959 from Dr Anderson to the author. One example of Anderson's sketch maps from Guatemala is given in figure 2.3.

28 Sauer, 'The education of a geographer' (note 17); quotation from pp. 294–6.

29 This view was anticipated more than a decade ago by H. Greenaway and H. Williams, *Patterns of Change in Graduate Employment*. London: Society for Research into Higher Education, 1973. For a picture of the jobs open to geographers, see J. Weltner and G. McBoyle, *Job Opportunities for Geography Graduates*, 6th edn. Waterloo: Department of Geography, University of Waterloo, 1988, p. 73.

30 The full version, 'To my wife. In memory of Bullock-cart days and Irrawaddy nights' was used by Dudley Stamp on the dedication page of *Asia: A Regional and Economic Geography*, 1st edn. London: Methuen, 1929. Stamp went first to Burma to explore for oil and was made the first professor of geography and geology in the University of Rangoon while still in his twenties.

31 See for example the multi-dimensional mapping programmes described in chapter 3. See figures 3.2 and 3.3.

32 A. D. Cliff and P. Haggett, 'Disease diffusion'. In Michael Pacione, *Medical Geography: Progress and Prospect*. London: Croom Helm, 1986, pp. 84–125.

33 Rosalind Franklin was an X-ray crystallographer who worked on the structure

of DNA with Maurice Wilkins at King's College, London, while James Watson and Francis Crick were working in the Cavendish Laboratory, Cambridge. She died from cancer in 1958 at the age of thirty-seven. Crick, Watson and Wilkins were awarded the Nobel Prize in 1962. The comment is attributed to Franklin in the BBC Horizon film *Life Story* based on the book by James D. Watson, *The Double Helix*. London: Weidenfeld and Nicolson, 1968.

Index